T0136909

ONE SHOT FOR GOLD

Mining and Society Series
Eric Nystrom, Arizona State University
Series Editor

"If it can't be grown, it must be mined," as the bumper sticker reminds us. Attempting to understand the material basis of our modern culture requires an understanding of those materials in their raw state and the human effort needed to wrest them from the earth and transform them into goods. Mining thus stands at the center of important historical and contemporary questions about labor, environment, race, culture, and technology, which makes it a fruitful perspective from which to pursue meaningful inquiry at scales from local to global.

Books published in the series examine the effects of mining on society in the broadest sense. The series covers all forms of mining in all places and times, building from existing press strengths in mining in the American West to encompass comparative, transnational, and international topics. By not limiting its geographic scope to a single region or product, the series helps scholars forge connections between mining practices and individual sites, moving toward broader analyses of the global mining industry in its full historical and global contexts.

Seeing Underground: Maps, Models, and Mining Engineering in America
Eric C. Nystrom

Historical Archaeology in the Cortez Mining District: Under the Nevada Giant
Erich Obermayr and Robert W. McQueen

The City That Ate Itself: Butte, Montana and Its Expanding Pit
Brian James Leech

Mining the Borderlands: Industry, Capital, and the Emergence of Engineers in the Southwest
Sarah E. M. Grossman

One Shot for Gold: Developing a Modern Mine in Northern California
Eleanor Herz Swent

ONE SHOT FOR GOLD

Developing a Modern Mine
in Northern California

ELEANOR HERZ SWENT

With a foreword by Eric C. Nystrom

UNIVERSITY OF NEVADA PRESS *Reno & Las Vegas*

Mining and Society
Series Editor: Eric C. Nystrom

University of Nevada Press, Reno, Nevada 89557 USA

LIBRARY OF CONGRESS CATALOGING-IN-PUBLICATION DATA
Names: Swent, Eleanor H. (Eleanor Herz), author. | Nystrom, Eric Charles, writer of foreword.
Title: One shot for gold : developing a modern mine in Northern California / Eleanor Herz Swent ; with a foreword by Eric C. Nystrom.
Other titles: Mining and society series.
Description: Reno ; Las Vegas : University of Nevada Press, [2021] | Series: Mining and society series | Includes bibliographical references and index. | Summary: "One Shot for Gold: Developing a Modern Mine in Northern California is an oral history of the development and operation of the McLaughlin Mine, the last major gold mine operated by the Homestake Mining Company in Napa County. It is a major contribution to mining history after 1980 in that it also depicts the lives of modern miners in a most productive gold mine of the late 20th century to the turn of the 21st"— Provided by publisher.
Identifiers: LCCN 2020051247 (print) | LCCN 2020051248 (ebook) | ISBN 9781647790066 (hardcover) | ISBN 9781647790073 (ebook)
Subjects: LCSH: McLaughlin Gold Mine (Calif.)—History. | Gold mines and mining—California, Northern—History. | Homestake Mining Company—History.
Classification: LCC TN413. Z6 M36 2021 (print) | LCC TN413. Z6 (ebook) | DDC 338.7/62234220979417—dc23
LC record available at https://lccn.loc.gov/2020051247
LC ebook record available at https://lccn.loc.gov/2020051248

First Printing
Manufactured in the United States of America

25 24 23 22 21 5 4 3 2 1

IN MEMORIAM

H. Rex Guinivere,
October 1, 1931–September 8, 2010
(*San Francisco Chronicle*, September 16, 2010)

James William Wilder,
April 4, 1924–December 17, 2010
(*Knoxville News Sentinel*, December 19, 2010)

William Albert Humphrey,
January 12, 1927–April 5, 2011
(*East Bay Times*, April 8, 2011)

John Sealy Livermore,
April 16, 1918–February 7, 2013
(*Napa Valley Register*, February 14, 2013)

Sylvia Cranmer McLaughlin,
December 24, 1916–January 19, 2016
(*San Francisco Chronicle*, January 21, 2016)

Donald Lee Gustafson,
July 8, 1938–May 24, 2017
(*Reno Journal-Gazette*, June 4, 2017)

CONTENTS

Part 4

Part 5

ILLUSTRATIONS

Maps and Figures (*following page 77*)

FOREWORD

Oral History and Mining History

Eric C. Nystrom

There are a number of places in the United States that would seem to be unlikely hosts for an open pit gold mine, but environmentalist California, in the same Napa County whose vineyards make world-class cabernet, might be among the least likely. Thus, the McLaughlin Mine was an idea seemingly incongruous with its context from the beginning. And that beginning had been in the 1980s, after the advent of environmental regulations had turned the domestic American mining industry upside down, forcing a reckoning with public perception and old habits. But what might be most remarkable of all is that the story of this mine—how it came to be, as told by the people who were there, on every side of the issue—was captured for posterity by a major oral history effort that was determined to save a full history of this place for the future.

More than a decade ago, I had the good fortune to meet Eleanor Swent at the Mining History Association. I got to know her over the years as a very modest and humble person, who would mention once in a while that she had done "a few" oral histories, hinting that it might be nice if someone could use them some day for a good book... and then, at one point, she revealed the full scope of what she had collected. In more than fifty interviews, her project had documented an open pit gold mine, in California, from start to finish.

The scope and significance of her collected interviews was stunning. Swent and her colleagues had tried to interview as wide a spectrum of people as possible, and not just those in favor of the project or those who worked for the mine. The interviews covered the entire history of the site, from before exploration, through discovery, development, environmental permitting, technological innovation, mining, and closure. She had done nearly all the interviewing herself, and every interview had been professionally transcribed. The rich content, inclusive scope, and professional execution of the project make the collection an invaluable treasure and probably unique.

And although mining historians have had to wait some two decades since the oral history documentation concluded, I think Eleanor Swent was right: It *did* add up to a good book.

One Shot for Gold sits at the juncture of modern mining history and oral history, and is an important contribution to our understanding of resource extraction in the late twentieth century. As a post–National Environmental Policy Act story of a new mining operation, in California no less, and based on interviews with people involved at every level of the project, the venture addresses a significant and under-documented period in mining history. As a well-planned and judiciously executed collective oral history, Swent's book provides a clear model that could, and should, be followed by future mining historians.

Academic historians will immediately notice that *One Shot for Gold* differs from the more traditional monographs that have been published in the University of Nevada Press's *Mining & Society* series thus far. For example, Swent's biography suggests an insider status, the text is lightly footnoted, and extensive quoted extracts are found throughout. It doesn't tackle big historiographical questions. Considered as a scholarly, collective oral history, however—in the spirit of Studs Terkel's *Hard Times*—these features are advantages or artifacts of the genre. The broader oral history project took place under the auspices of University of California, Berkeley's (UC Berkeley) esteemed Regional Oral History Office, which has been helping to set the standard for scholarly practice of oral history for more than half a century. Swent's personal familiarity with the people, places, and technology of mining helped the project gain credibility, and put her interviewees at ease. Readers don't immediately see Swent's hours of preparation for each interview, reading up on the technical details of the mining properties and processes to be discussed, but that rigorous foundation work is immediately apparent in the quality and depth of the interviews. The text permits the people being interviewed, or narrators, wide latitude to speak for themselves, to be whole personalities rather than just evidence for an author's point of view.

Broadly, it has long been common for many people associated with mining to be defensive, even to the point of appearing reactionary, in discussions about negative impacts associated with their industry. As interviewer and insider, Swent is able to move beyond these positive/negative dualities to encourage her narrators to portray a more complete, authentic picture. This project isn't a cartoonish mine, all smiling men and women in hardhats, nor is it a malevolent beast conjured by earth-destroyers. Instead, the people interviewed largely, though not exclusively, thought they were doing a good thing. They worked very hard, relative to the standards of the day, to do a mine the "right" way, within a pragmatic framework. Their efforts tried to make Homestake a good neighbor to Lake, Napa, and Yolo Counties and to leave a positive environmental legacy while keeping in mind the fundamental truth that mining was a business, intended to extract minerals from a specific bit of the earth for a determinate length of time in order to make a profit. It's complicated, and we get to see that.

In addition to the coverage and the nuanced sensitivity, Swent's oral histories stand out because of her comfort as an interviewer with the technical details of modern mining practice. The details here are, in my view, rather important, since there are very few detailed portraits of this sort of place in the field of mining history, and it's difficult to get to those details from other sources. Indeed, a significant source of the value in this project is the richness of the portrait that can be drawn, since it is a time and place—California, late twentieth century, regulatory era, new project, large company—that can be very difficult to accurately depict at all. Furthermore, Swent was able to interview up and down the Homestake corporate hierarchy and inside and outside the company. Of the studies in mining history we do have of the recent era, many of them—while quite good—have focused on the executive-level corporate perspective, making it difficult to see details from the ground-level viewpoint.[1]

Academic practitioners of oral history usually date their field's roots to the late 1940s and early 1950s, when the first oral history projects were established at universities such as Columbia University and UC Berkeley. However, mining historians might fruitfully suggest that the oral history

1. Examples include Duane A. Smith, *Staking a Claim in History: The Evolution of Homestake Mining Company* (Walnut Creek, CA: Homestake Mining Company, 2001); Jack H. Morris, *Going for Gold: The History of Newmont Mining Corporation* (Tuscaloosa: University of Alabama Press, 2010); and Charles Caldwell Hawley, *A Kennecott Story: Three Mines, Four Men, and One Hundred Years, 1897–1997* (Salt Lake City: University of Utah Press, 2014).

of mining was being done long before it found a home in mid-century academic halls.

Perhaps the most significant of the early mining oral historians was Thomas A. Rickard. Rickard was a trained mining engineer from a prominent British mining family who made his greatest impact on the field through his writing and editing activities. He served at one time or another as editor and/or publisher of the greatest industry publications of the early twentieth century, including the *Engineering and Mining Journal*, the *Mining and Scientific Press*, and the *Mining Magazine* of London.[2] Late in life he made several efforts to write the mining industry's history,[3] but in the first decades of the twentieth century—before tape recording was common—he conducted what is recognizable today as a mining oral history project. In 1922 he published a set of personal interviews he had conducted with leaders of the mining industry, beginning in 1915.[4] Anticipating the modern oral history form, these interviews were structured so that Rickard's questions and his subjects' answers were made distinct in the text, and were conveyed in lightly edited spoken language. This set of twenty-three interviews—plus one more placing Rickard himself on the hot seat—covers more than five hundred pages and permits the modern reader to hear the words firsthand from prominent engineers whose careers began as early as the late 1860s.

The mining industry was one of the topics of interest for early academic oral history programs as well. The beginnings of modern oral history collections at UC Berkeley occurred in the early 1950s under the direction of Corinne Gilb, a pioneer of oral history methods who founded and administered the Regional Cultural Oral History Project, which later became famous in academic circles as the Regional Oral History Office (and today is the Oral History Center).[5] Gilb conducted interviews with a number of

2. "T. A. Rickard (Deceased 1953)," American Institute of Mining Engineers, http://www.aimehq.org/programs/award/bio/t-rickard-deceased-1953 (accessed Dec. 12, 2019).

3. T. A. Rickard, *A History of American Mining* (New York: McGraw-Hill, 1932); T. A. Rickard, *Man and Metals: A History of Mining in Relation to the Development of Civilization* (New York: McGraw-Hill, 1932).

4. T. A. Rickard, *Interviews with Mining Engineers* (San Francisco: Mining and Scientific Press, 1922), available from https://archive.org/details/interviewswithmioorickuoft. The earliest interview was that of Charles Butters, originally published in the *Mining and Scientific Press* of August 21, 1915.

5. "Corinne Lathrop Gilb: Oral Historian," http://corinnelathropgilb.com/oral-historian.html; "Finding Aid to the Corinne Gilb Collection," Labor Archives and

mining history–related narrators, from a group of old-timers in the mining ghost town of Hornitos to William E. Colby, a mining lawyer and Sierra Club secretary.[6] The Columbia Center for Oral History in New York sponsored an effort in 1960–61 by mining engineer Henry C. Carlisle to interview many of his fellow mining engineers who were then in the twilight of their lives. The twenty transcribed interviews that resulted (Carlisle, like Rickard, also recorded his own history) provide an important set of firsthand accounts documenting mining engineering from the 1910s or 1920s through World War II.[7]

From the 1970s onward oral history became an essential part of any social historian's toolkit, and thus an indispensable technique for writing modern history. Oral historians sought out and interviewed working-class people associated with mining, in addition to those such as mining engineers who made appearances in earlier efforts. A handful of mining places attracted the attention of oral historians for many decades, such as Butte, Montana, and the coalfields of Appalachia.[8] Others were chronicled in individual documentation projects or by a lone interested historian.[9] Many other interviews rest in archives and personal collections, awaiting future historians to use them to interpret the mining past.

Research Center, San Francisco State University, https://oac.cdlib.org/findaid/ark:/13030/c8cf9w68/.

6. "Oldtimers of Hornitos," *Life in a Mining Town: Hornitos, California*, interviewed by Thomas Coakley and Corinne Gilb in 1954 (Berkeley: Regional Cultural History Project, University of California Library, 1956), https://digitalassets.lib.berkeley.edu/roho/ucb/text/hornitos.pdf; William E. Colby, *William Edward Colby: Reminiscences*, interviewed by Corinne Gilb in 1953 (Berkeley: Regional Cultural History Project, University of California Library, 1954), https://archive.org/details/remwilliamedwardoocolbrich.

7. "Mining Engineers Project: Oral History, 1961," Columbia Center for Oral History, Columbia University Libraries, New York, https://oralhistoryportal.library.columbia.edu/document.php?id=ldpd_4076593.

8. Mary Murphy, *Mining Cultures: Men, Women, and Leisure in Butte, 1914–41* (Urbana: University of Illinois Press, 1997); Brian James Leech, *The City That Ate Itself: Butte, Montana and Its Expanding Berkeley Pit* (Reno: University of Nevada Press, 2018); Alessandro Portelli, *They Say in Harlan County: An Oral History* (New York: Oxford University Press, 2011).

9. Many such examples might be noted here, but particularly intriguing are those where the impetus, funding, or both were supplied by the government or mining firms as a means of cultural preservation in the context of permitting processes for new development. See, e.g., Dana R. Bennett, *A Century of Enthusiasm: Midas, Nevada, 1907–2007* (Midas, NV: Friends of Midas, 2007); and Victoria Ford, *Silver Peak: Never a Ghost Town* (Reno: University of Nevada Oral History Program, 2002).

But those future generations of mining historians will surely be familiar with Eleanor Swent, since she was the driving force and primary interviewer for the *Western Mining in the Twentieth Century* oral history series, conducted for the Regional Oral History Office of UC Berkeley. Between the mid-1980s and the early 2000s, Swent conducted more than one hundred interviews of people associated with the mining industry. An active and well-connected advisory committee marshalled support from a wide swath of the hard-rock mining industry, especially that portion related in some fashion to California, and a large grant from the Hearst Foundation (created with a fortune that started with William Randolph Hearst's father George, a nineteenth century mining magnate who owned the original Homestake Mine) endowed the work.

The project's beginnings were at once ambitious and modest. Professor Douglas Furstenau with UC Berkeley, and Eleanor Swent and her husband Langan, together drove up to the dedication of Homestake's new gold mine in California, and had some time to ruminate during the drive. How often does a historic moment become clear only in retrospect, once so many of the people involved can no longer contribute to depicting its history? During that drive the group hatched the idea to document a modern mine's complete lifecycle from opening to closing—*this* mine, the McLaughlin Mine in Napa County, California. This was one of the first attempts at a new full-scale mine by a major corporation in California after the advent of environmental regulations, and the first full-scale use of autoclaves to extract gold from ore. Why not interview a wide cross section, from executives to geologists to engineers to community members, to capture a sense of what modern mining was really like?

Eleanor Swent was an ideal project director and interviewer. Daughter of the chief metallurgist at the famous Homestake Mine, she had grown up during the Great Depression in the mining town of Lead, South Dakota, before earning degrees in writing at Wellesley and the University of Denver. After she married Langan Swent, a mining engineer, they lived in mining towns in Mexico, South Dakota, and New Mexico, and eventually settled in Northern California because of his work at Homestake Mining Company's San Francisco headquarters, where, in the late 1980s, Langan Swent neared retirement. Almost from their arrival in California in the late 1960s, Eleanor Swent taught English as a second language part time to adults through the public-school system. The accounts of some of the refugees who appeared in her classroom fascinated her, so she began recording their stories—with a little helpful advice from Willa Baum who, like Swent, had taught adults in Chinatown years before and who had succeeded Gilb as longtime director of the Berkeley Regional Oral History Office.[10]

The effort to document the McLaughlin Mine and the Knoxville district ultimately produced a mining oral history collection of unparalleled diversity and depth, printed in eight volumes of interviews, and supported by many other stand-alone interviews as part of the broader *Western Mining in the Twentieth Century* series. The first volume, containing four interviews, suggests the range of people consulted throughout the collection: a mining company vice president, a citizen activist, a metallurgical technician, and the director of the local water district.[11] Homestake employees made up a substantial number of interviews, from executives, geologists, and engineers who played important roles in discovering, planning, operating, financing, and ultimately shutting down the mine, through to some employees who might have been easily overlooked. Neighbors of the mine, officials from county government, townspeople, and activists were represented. Nearly a quarter of the interviewees in the eight volumes were women.[12]

The aim was to document the mine from its very beginning to its eventual end and transformation into a designated nature reserve. This puts a distinctive twist on the selection of interviewees. Though some mine workers are represented, the focus is not on the day-to-day work of an operating mine. Instead, the goal was to capture the full context of how the mine came to be and what it meant. This required attention to multiple levels, and to different types of people. It was as essential to document the mindset of Homestake's executive decision-makers, as it was to document the mindset of the geologists exploring the land, or of the county officials skeptical of the proposal, or of the landowner who sold the company the mine. Decisions were made by humans at every level, interacting with the

10. Eleanor Herz Swent, *One Woman's Experience with the Role of Women in the Mining Community*, interview by Malca Chall in 1994 (Berkeley: Regional Oral History Office, The Bancroft Library, University of California, 1995), 980–89, https://archive.org/details/safetyinundergroo2swenrich/page/895. (Note that the transcript of the interview with Eleanor Swent is appended to the end of volume 2 of interview of Langan Swent and is not paginated separately.)

11. Knoxville Mining District, *The McLaughlin Gold Mine, Northern California, 1978–1995*, vol. 1, *An Oral History Conducted in 1994 and 1995*, Regional Oral History Office, The Bancroft Library, University of California, Berkeley, 1998, https://archive.org/details/knoxvilleminingo1swenrich. Interviews are in alphabetical order.

12. Ten out of forty-one interviewees across the eight collected volumes were women.

choices and contexts formed by the people around them. Out of all of this, balancing interests with costs and impacts, emerged a mine.

Swent and the *Western Mining* oral history project did not stop with the completion of the McLaughlin Mine interviews. She continued her work with the project into the 2000s, eventually creating more than a hundred interviews—a priceless body of knowledge about twentieth-century mining on which future historians can draw. But the masterpiece of this work is undoubtedly the McLaughlin project, and we are fortunate that Swent has knitted together the fascinating history of this project from the rich interviews she collected.

One Shot for Gold is an important story—the full story of the first modern gold mine, in the environmental regulation era, in California. The oral histories it is based on were conducted according to the highest modern standards, with extensive planning; careful selection of narrators to ensure a range of expertise, positions, and viewpoints; multiple interviews to capture the entire story; and careful editing and transcription, approved by the narrator, to ensure accuracy. The end result is a conscious co-creation of an oral document that is intended, by both narrator and interviewer, to be a lasting historical resource. On a foundation constructed with such depth and sensitivity, the story of the McLaughlin Mine effectively poses juxtapositions that spark the interest of readers approaching the topic from any side.

People associated with the modern mining industry, especially in environmental permitting, remediation, and the Corporate Social Responsibility (CSR) sectors, can see some of the early precedents in those areas, and other audiences beyond historians will find much to enjoy in this readable account. The broader goal, as with any good work of history, is to help broaden understanding. This was clear to Eleanor Swent from the start. When asked to reflect on the oral history project and what might come of all the interviews she had conducted, she replied, "[I] hope that these will be used for research by people. The ones I interviewed already know how important it is but I hope that they'll be used by a wider audience who might appreciate these people."[13]

— February 19, 2020

13. Eleanor Swent, "An Oral History: Eleanor Swent," conducted by Paul Burnett in 2013, Regional Oral History Office, The Bancroft Library, University of California, Berkeley, 2013, p. 24.

PREFACE

Wherever there is a hope that pure gold or gems may be found, the ground can be turned up.

— Georgius Agricola (Georg Bauer), *De Re Metallica, 1556*
(Translated from the German by Herbert Clark Hoover
and Lou Henry Hoover)

In 1978 Donald Gustafson, a young geologist working for Homestake Mining Company, California's oldest corporation, discovered gold at the site of a historic mercury mine about an hour's drive north of San Francisco. Lawyers working from the company headquarters, occupying several floors of a high-rise building in San Francisco's financial district, began negotiations with James William Wilder, the owner of the One Shot Mining Company, in his office in a trailer without a telephone, on an unpaved road southeast of the town of Lower Lake. The mine that was eventually developed, named in honor of Homestake chairman emeritus, Donald Hamilton McLaughlin, was California's most productive gold mine of the twentieth century. Between 1985 and 2002 the mine produced about 3.4 million ounces of gold. Homestake Mining Company's first gold mine was discovered in 1876, almost exactly a hundred years earlier, in the Black Hills of South Dakota, and it was still operating, although nearing its end. Now the company was challenged to develop a mine at a very different time and place.

At the same time that Don Gustafson was searching for a gold mine, Americans were being energized by a new idea—environmental protection—and this movement was particularly strong near San Francisco. In 1961 Sylvia Cranmer McLaughlin, Donald McLaughlin's wife, was a founder of Save San Francisco Bay, one of the first grassroots organizations of its kind.

The first Earth Day was celebrated in San Francisco in 1970 and attracted enthusiastic young people from all over the country. They did not know that gold would be a critical component of their new computers and cell phones. They only knew that black lung disease killed coal miners and they sang along with Tennessee Ernie Ford,

You load sixteen tons, what do you get?
Another day older and deeper in debt.

There was still another trend for Homestake to consider: a nation that had celebrated beer and whiskey now enjoyed wine, and Napa Valley wineries were gaining worldwide recognition and attracting tourists. And now—a mine—in Napa County? Few people knew that the mine site, while technically in Napa County, is separated from the vineyards by miles of forest-covered mountains.

Homestake succeeded in developing a mine that protected the environment and is acclaimed as a model for mine reclamation. The management style was distinguished by a sense of teamwork and mutual respect at all levels, and all activities were exceptionally transparent and open to the public. The outcome, far from being assured, was the result of many patient negotiations and problems solved. Some were technical; many were political. Homestake personnel had to face public ignorance and fear as well as a daunting number of bureaucratic requirements. The mine is near the junction of three counties: Lake, Napa, and Yolo; and there were multiple federal, state, and local agencies with jurisdiction over air, water, roads, and dams. Protection of the environment had to be assured to each of them, in order to obtain an unprecedented number of 327 permits. To satisfy the requirements, a revolutionary process to recover the gold by high-pressure oxidation, emitting no pollutants of air or water, was developed that has been adopted around the world.

The mine site was accessible from the town of Lower Lake in Lake County. A century earlier, in 1862, the county's thriving town of Knoxville had a store, hotel, post office, and Wells Fargo office. The mercury mines there provided the financial basis for families that later were known for their wealth and/or political power. Because of its importance, in 1872, Napa County, with its larger population, paid Lake County $3,500 to annex the Knoxville Township. In 1970 Lake County had become the poorest county in California, with a median household income of $5,266, and a population of 19,548. By 1989, when the gold mine was in production, the median household income had about quadrupled, to $21,794. The population in 1990 had more than doubled, to 50,631; by 2000 it had grown to 55,000. With funding support from Homestake, a community college branch was subsidized to train local workers, a hospital expanded its services, telephone and electric utility service was extended, and roads were paved. Many of these improvements provided lasting benefits to the county.

The mine operations ceased in 2003, and the reclaimed site is now the Donald and Sylvia McLaughlin Natural Reserve, part of the University of California system, where scholars conduct research on the unique serpentine ecology. Donald Hamilton McLaughlin was the protégé of Phoebe Hearst, widow of the founder of Homestake Mining Company. In 1926 he became consulting geologist for the company; from 1943 he was successively director, president and chief executive, chairman, and chairman emeritus of Homestake. He graduated from the University of California, Berkeley (UC Berkeley) in 1914; in 1941 he returned as professor of mining engineering and then dean of engineering. In 1951 he was appointed a regent of the University of California. He died at the age of ninety-three on December 31, 1984, and Sylvia McLaughlin dedicated the mine to him in a ceremony on Saturday, September 28, 1985. The natural reserve named for them is a fitting coda to the story of a modern mine: extracting one precious resource, gold, and preserving another, the natural environment, its air, and its water.

This was also the final chapter in the history of Homestake Mining Company, which held its last shareholders meeting on December 14, 2001.

In January 1986 the Regional Oral History Office, now the Oral History Center, at UC Berkeley, in line with its mandate to document the lives of significant figures in the development of northern California, the West, and the nation, established the oral history series on Western Mining in the Twentieth Century (WMTC) to document the lives of leaders in mining, metallurgy, geology, education in the earth and materials sciences, mining law, and the pertinent government bodies. The field includes metal, nonmetal, and industrial minerals. Mining has changed greatly in this century: in the technology and technical education, in the organization of corporations, in the perception of the national strategic importance of minerals, in the labor movement, and in consideration of health and environmental effects of mining. The idea of an oral history series to document these developments in twentieth-century mining had been on the drawing board of the office for more than twenty years. The project finally got under way on January 25, 1986, when the director of the office Mrs. Willa Baum, Mr. and Mrs. Philip Bradley, Professor and Mrs. Douglas Fuerstenau, Mr. and Mrs. Clifford Heimbucher, Mrs. Donald McLaughlin, and Mr. and Mrs. Langan Swent met at the Swent home to plan the project, and Professor Fuerstenau agreed to serve as principal investigator. The project was presented to the San Francisco section of the American Institute of Mining, Metallurgical, and Petroleum Engineers (AIME) on Old-Timers Night, March 10, 1986, when Philip Read Bradley Jr. was the speaker. This section and the Southern California section of AIME provided initial funding and organizational sponsorship. The Northern and Southern California sections of the Woman's Auxiliary to the AIME (WAAIME), the California Mining Association, and the Mining and Metallurgical Society of America (MMSA) were early supporters. Later the National Mining Association became a sponsor.

After more than thirty individual biographical oral histories were completed, a community oral history was undertaken, documenting the development of the McLaughlin gold mine in the Napa, Yolo, and Lake Counties of California (the historic Knoxville mercury mining district), and the resulting changes in the surrounding communities. Future researchers will turn to these oral histories to learn how decisions were made that led to changes in mining engineering education, corporate structures, and technology, as well as public policy regarding minerals.

In 1991 the special Knoxville District/McLaughlin Mine oral history project was launched. The life of the mine was projected to be about twenty years, and most of the key players were available for interviews. It was a rare opportunity to document the discovery, development, and closing of a mine while it was happening. The Knoxville/McLaughlin project includes forty-eight interviews, from two to seventeen hours long, conducted with Homestake officials and employees, a supervisor from each of the three counties involved, Napa County planners, the Lake County school superintendent, community historians, mercury miners, merchants, and ranchers, as well as some of the most vocal opponents of the mine. Their voices help to tell the story of the McLaughlin Mine.

> Oral history provides a fuller, more accurate picture of the past by augmenting the information provided by...other historical materials...[and] helps us understand how individuals and communities experienced the forces of history.

> Eyewitnesses to events contribute various viewpoints and perspectives that fill in the gaps in documented history, sometimes correcting or even contradicting the written record. (Baylor University 2012, n.p.)

In the final interview, recorded in July 2001, the McLaughlin Mine's environmental manager Raymond Krauss tells of the fruition of his vision: negotiating a donation of the last bit of land that Homestake owned to the American Land Conservancy.

The interviews were conducted by Eleanor Herz Swent, a past president of the Mining History Association and winner of the Rodman Paul award for contribution to mining history. She was born and raised in Lead, South Dakota, the home of the Homestake Mine, and lived for many years in mining communities. Her mother had a geology degree; her father, Nathaniel Herz; husband, Langan Swent; and father-in-law, James Swent, were all mining engineers with professional connections to Homestake.

Duane A. Smith, professor of history and Southwest studies at Fort Lewis College, Durango, Colorado, in the introduction to the oral history series, says,

> This multi-volume series covers almost every conceivable aspect and impact—it is a monument to a refreshing, innovative way of approaching mining history.
>
> These volumes provide a case study of twentieth century mining, environmental issues, and regional concerns, the successes, failures, tensions, and developments that go to make up a 1980s and 1990s mining operation and the people involved from all walks of life. They are a gold mine of primary documentation and personal memories of an era. (Knoxville Mining District [hereafter KMD] vol. 1, n.p.)

ACKNOWLEDGMENTS

Mining has been important in the history of California since before it was a state, and Willa Baum, director of the Regional Oral History Office (now the Oral History Center) at the Bancroft Library of UC Berkeley, for many years had hoped to document developments in twentieth-century mining. When we met in 1974, I was a volunteer and amateur interviewer for an Oakland Neighborhood History project funded by the National Endowment for the

Humanities and Willa Baum was an advisor to the project. As we became acquainted, she said that, with my interest in oral history and familiarity with mining, a mining history project could be developed. This was years before the discovery of the mine that is the subject of this book. Mrs. Baum urged me to take Mark Greenside's class on oral history methods at Merritt College and Charles Morrissey's oral history workshop in San Francisco, and to join the Oral History Association.

I was hired in 1986 as director of the Western Mining in the Twentieth Century series. The first interview I conducted was with Horace Albright, UC Berkeley class of 1912, who said, "I was educated as a mining lawyer, and you get mining in your blood" (Albright 1986, xv). My husband, Langan Swent, until his death in 1992, delighted in referring to himself as chief technical advisor to the series; he helped me with his vision, patience, and wise counsel. William G. Langston, Homestake vice president and general counsel from 1973 to 1992, provided background information and encouragement from the beginning. Professor Douglas Fuerstenau agreed to serve as principal investigator for the mining series; he provided encouragement and practical help for more than three decades, and continues to do so.

Members of the Mining History Association, especially Eric Nystrom, Duane Smith, and Sally Zanjani, encouraged the mining series and the subsequent Knoxville/McLaughlin project. I cannot adequately thank Eric for his thoughtful foreword to this manuscript and Dean Enderlin for the appendix tour guide script..

Thanks are due to all of the interviewees who gave their time and energy to recording their stories, and especially to Homestake employees Dean Enderlin and Raymond Krauss, who were instrumental in facilitating other interviews and providing source information; William Humphrey, who secured significant funding from the Hearst Foundation; Dennis Goldstein for his encouragement; and Jack Thompson, who provided background material and photos from his files. Suzanne Riess, a senior editor at the Oral History Center, made an invaluable film, narrated by Patrick Purtell, of the McLaughlin process plant. Her film is in the Bancroft Library archives.

Others at the Oral History Center were instrumental: Julie Allen, David Dunham, and Martin Meeker, who later became director of the Oral History Center. Laura McCreery, a former editor at the OHC, gave welcome advice and encouragement.

Cathy Koehler, director of the Donald and Sylvia McLaughlin Nature Reserve, provided overnight hospitality for Sylvia McLaughlin and me on

April 24–25, 2007, at the former warehouse, now the Ray Krauss Field Station, and led a tour of the reserve. Cathy generously offered information from her files in subsequent years.

John Livermore gave a tour of the Corona and Oat Hill Mine properties and much of the back country of Napa County, and introduced me to Tony Cerar. He also facilitated the Hugh Ingle interview and oral history.

Members of the Lake County community helped: local historian Kevin Engle; Lake County library technician Jan Cook; online editor Samie Hartley and archivist Sean Scully of the *Napa Valley Register*; Elizabeth Larson of the *Lake County News*; and Linda Lake at the Lakeport Museum. Margaret Vance generously put me up at her home on the shore of Clear Lake, a delightful respite after long hours of interviewing.

Those who painstakingly transcribed the interviews from the tapes were Shannon Page, Mim Eisenberg/WordCraft, Julie Allen, Teresa Bergen, Aric Chen, Lisa Delgadillo, Melanie Schow, Estevan Sifuentes, Gary Varney, and Lisa M. Vasquez. Final typists were Amelia Archer, Shana Chen, Merrilee Proffitt, and Jennifer Thoms. Thanks to them all, especially to Mim Eisenberg, who attached a note saying she felt better about mining because of Homestake's care for the environment.

Members of the staff at the Oral History Center at Berkeley have continued to be helpful: Martin Meeker, director; Paul Burnett, historian/interviewer; and David Dunham, operations manager. Laura McCreery, former colleague at the Oral History Center, reviewed the manuscript and gave constructive advice. Theresa Salazar, curator, Western Americana; and Peter Hanff, deputy director at the Bancroft Library, helped with access to archives there, and introduced me to Phil Hoehn, who has helped at the Rumsey map center at Stanford. Lauren Lassleben, appraisal and accessioning archivist at the Bancroft Library, was also helpful. Gray Brechin, fellow member of the California Studies Group at UC Berkeley, has given advice and encouragement.

Thanks are owed to Ed and Gretchen Hillard (introduced to me by my son Richard and daughter Jeannette) and Juli Armstrong (introduced by Jean Doble) for personal encouragement as well as technical aid; to Robert Anderson (introduced by Betty Kaplan) for final manuscript formatting; and to Tim Dec for computer help. I am not allowed to know who reviewed the manuscript for the University of Nevada press and approved it for publication, but I am grateful to them, whoever they are. Margaret Dalrymple, acquisitions editor at the University of Nevada Press; JoAnne Banducci, director; Sara Vélez Mallea, project manager; and Alison Hope, copy editor, have been considerate and patient with my shortcomings.

Thanks to my children and grandchildren who have always inspired and encouraged me.

Please forgive any omission from this list.

PART 1

CHAPTER 1

MERCURY

1. A Roman deity, the god of eloquence, skill, trading and thievery
2. The heavy silver-white liquid metal otherwise called quicksilver. It absorbs other metals, forming amalgams, and is commonly obtained by sublimation from cinnabar, its most important ore.

—*Oxford Universal Dictionary* (1955)

Scientists at the National Aeronautics and Space Administration (NASA) tell us that about four and a half billion years ago, gravity pulled swirling gas and dust together and formed our planet Earth, a molten ball of minerals whirling through space. As it cooled, the center remained hot, about 6,000 degrees Celsius, but a cooler surface crust formed, floating on the magma, and in time breaking into continent-size pieces. One of those pieces is the North American tectonic plate, and next to it is the Pacific plate, dipping under the Pacific Ocean. As the plates shifted, the Pacific plate moved under the North American plate, tipping up the coast of what would one day be called California. Mountain ranges formed, and volcanoes and fissures spewed steaming lava. This action continues today, causing disastrous earthquakes, as well as hot springs that relax and heal sore bodies, and fumaroles that can be tapped to produce electric power. Two important minerals rose from the boiling center of the planet: mercury and gold. The mercury was deposited in visible bright red veins spread throughout the crust. As for the gold, it was only in the late twentieth century that geologists developed and tested the theory that this precious metal, too fine to be seen with the naked eye, could be found in hot springs as well as in visible veins.

In 1849 miners came from around the world to hunt for gold in the streams that flowed from the mountains that the Spanish called Snowy Mountains (in Spanish, Sierra Nevada) of California. At first the prospectors looked for gold nuggets—placer gold—in the streams, and then they went upstream and dug gold-bearing ore from the veins in the hard rock there. They ground the rocks into small grains, and turned to an ancient process, amalgamation with mercury, to free the gold.

Gold is known to metallurgists as a noble metal because it does not oxidize or combine easily with other metals. Mercury's ability to mix with gold so that it can be extracted from ore has always created a demand for mercury wherever gold is mined. At the gold mines in the Sierra Nevada, the gold ore was brought to the surface, crushed, and mixed with mercury. The amalgam (a soft mass, formed especially by combination with mercury) was then heated in a furnace. Mercury boils at 357 degrees Celsius (357°C) and gold melts at 1064°C and boils at 2836°C; heating the amalgam above mercury's boiling point drives off the mercury and leaves behind the gold. The mercury vapor can be condensed into liquid form and the mercury used again.

The search for mercury after 1850 led to the area between Clear Lake and the valley where a Basque family named Berrelleza had settled and claimed vast areas of ranchland, including what is now Lake Berryessa, an English spelling of the name. From then on, mercury mines, along with cattle ranching, sustained the economy there. By 1981 about 100 million kilograms of mercury had been produced in California. Ninety percent of the mercury mined in the United States has been mined in California, most of it from Lake County. It commonly occurs there in veins as cinnabar (mercury sulfide), which is prized by artists for its vermilion color. Sometimes it occurs as a silvery liquid that oozes from the rock. The droplets are hard to capture, so it is also called quicksilver. Mercury is used for mirrors, thermometers, medicines, and as a detonator for explosives, as well as for gold recovery. Since Biblical times, mercury has been measured by the flask, which is 34.47 kilograms or 76 pounds.

CHAPTER 2

KNOXVILLE AND MERCURY MINERS

We'll dig some more holes
Like the lowly moles

—Elmer Enderlin, miner and poet

The Hatter's remark seemed to have no sort of meaning in it, and yet it was certainly English.

—*Alice's Adventures in Wonderland,* Lewis Carroll

In France in the seventeenth century, felt hats came into vogue. After the felting process began to use mercury, hat makers suffered from the neurotoxic effects of exposure to mercury vapors. By the Victorian era, "mad as a hatter" was a common expression. Mercury in metallic form, however, is not toxic, and some of the men who worked in the mercury mines around Knoxville lived remarkably long and healthy lives.

Knoxville became Lake County's center for mercury mining in 1861 when Raynor and Richard Knox developed a deposit known variously as the Knoxville, the Redington, and the Excelsior, or XLCR Mine. This last name was ingeniously formed from the names of its owners: X for Knox; L for Horatio Gates Livermore; C for Caleb Hobbs; R for John Redington. Knox's name lived on in the Knoxville town and district; a century later, Horatio Livermore's great-grandson John Sealy Livermore would play a key role in a different kind of mining in the Knoxville area.

The Knoxville Mine was officially established in 1861, and it employed as many as three hundred men. The town of Knoxville had thirty or more buildings, including a store, hotel, Wells Fargo office, and school, as well as a cemetery. It had its own post office and Knox was postmaster. Because of its importance, in 1872, Napa County, with its larger population, paid Lake County $3,500 to annex the Knoxville Township.

The mercury mines had a variety of names, both descriptive and romantic: Abbott, Aetna, Bella Oak, Big Injun, Boston, Bradford, Corona, Great Western, Helen, La Joya, Mirabel, Oakville, Oat Hill, Pioneer, Reed, Wide-Awake, Red Elephant. The Manhattan Mine was opened in 1869 by R. F. Knox and Joseph Osborn on the same lode as the Redington Mine. In 1874 Knox and Osborn developed a successful and widely copied continuous-feed furnace for recovering the mercury. The district thrived until the beginning of the twentieth century, when California gold mining had declined from the peak years of the 1860s. Thereafter, demand for mercury fluctuated, and the fortunes of Knoxville rose and fell accordingly. There is always a demand for mercury for mirrors, paint color, thermometers, and medicines. In 1914 the beginning of World War I sparked a new demand for mercury as a detonator for explosives, and the mines of Lake County, including the Knoxville and the Manhattan, prospered again.

In 1930 the wealthy Gamble family of Palo Alto owned the Knoxville area. A century earlier, in 1830, James Gamble immigrated from Ireland to Cincinnati, Ohio, where he made soap. He and William Cooper Procter, a candle-maker from England, married sisters, Olivia and Elizabeth Norris, and in 1838 they opened a soap and candle factory. Successfully marketing "99.44% pure" Ivory soap, Procter & Gamble, or P&G, was listed on the New York Stock Exchange in 1891. Its enlightened management gave workers a half-day off on Saturday, a most unusual benefit at that time (Procter & Gamble n.d.).

In 1901 William Gamble's son Edwin came with his wife and four children to California to retire and built a home that is now a showplace in Palo Alto. He died on April 26, 1939. By then, the family had acquired ranching and wine-producing land in Napa County, including the Knoxville Mine, as well as the Kenton gold mine in Alleghany, California.

One of Edwin's three sons, George Gamble, graduated from Stanford University in 1908 as a mining engineer, and operated the family's mines. He was admired at Knoxville as a hard worker, drilling, blasting, and mucking out (or shoveling), along with the miners he employed. He was generous and paid above-average wages. Houses were built for the families of married men, and a bunkhouse and a good cook were provided for unmarried mine workers. They were given ample lunches: roast beef sandwiches, fresh apple pie, or chocolate cake. There was a schoolhouse that included living quarters for the teacher, who was always an unmarried woman. Some children from the nearby Reed Mine also attended the Knoxville school.

World War II brought a prosperous period for the mercury mines. On October 8, 1942, the War Production Board issued Order L-208 that shut down gold mines, since gold was not considered an essential mineral. Mercury, on the other hand, was in demand for use as a detonator in explosives, and Knoxville thrived again.

Edward McGinnis was born in Arbuckle, California, in 1923 to a pioneer family in the Berryessa Valley, now submerged in Lake Berryessa. His paternal grandfather, Noble Hamilton McGinnis, was born in Tennessee, and came to the Berryessa Valley with his wife after serving in the Civil War. A brother-in-law, F. E. Johnston, was already settled in the valley, later raising cattle on the property where the Knoxville Mine was located. When he was in his seventies, Edward McGinnis recalled that after graduating from high school in Winters in 1941, he went to work firing the retort at the Skaggs Springs mercury mine. In 1995 he recounted his experience:

> I'm talking about during World War II. And just before and afterwards.... About 1939 they started gearing up things for World War II, and they wanted that mercury, and they got bids for the federal government to come in and help Napa County with the road going from Monticello to Knoxville, because everything in that period come out through Napa County. [The] trucks hauled the mercury out to Sacramento...about twice a week. (McGinnis interview, KMD vol. 5, 240–41)

He worked at a retort, which he explains:

> In the old days they built it up out of fieldstone rock around a metal enclosure, then built a fire in the fire box...they put the ore in trays and put them in a big, like a bakery thing and it cooked the cinnabar into mercury. It went into vapor, then it would condense and went back to liquid.... The retorts, they couldn't run very much through a day with these. Reed Mine, Knoxville, and Manhattan, they all had the furnaces.... On the bigger operations they had a furnace that was on an incline. That heated up real hot, it was fired by oil, and they fed the ore into it and that kept revolving until all went through, and it cooked the mercury that went to vapor, and then they had the big condensers, and all that smoke went through, and mercury would run down into pots underneath these condensers...the Reed Mine and Knoxville I'd say there was probably, like, a hundred people in each place.

> That one mine I worked the retort…never worked that long around the furnaces…but at the Reed, I worked as a carpenter's helper. The first ones to work there really didn't know anything about it. They was people worked on them furnaces especially. They were the ones that got the fumes. Made their teeth fall out, their hair fall out. So it wasn't good. (McGinnis interview, KMD vol. 5, 238–42)

He left in 1943 to join the Army, and served with the 87th Mountain Infantry Regiment, later the 10th Mountain Division or the ski troops, first in the Aleutian Islands and then in Leadville, Colorado. He later worked for the Napa County highway department for more than thirty-six years.

Thornton Scribner, a Stanford University graduate and mining engineer, was superintendent of the Kenton gold mine in Alleghany, California; George Gamble was the mine's owner. Then the Kenton Mine shut down because of WPB Order L-208.

> The closure order ("Order L-208") had been issued by the War Production Board, a government entity created in 1941 to assist with converting America's civilian industry to wartime production. The War Production Board gave priority to copper mining, which had useful military implications, and labeled gold mines as "nonessential" for purposes of WWII. As such, Order L-208 prohibited owners of "nonessential" mines from taking any action to "acquire, consume, or use any material, facility, or equipment to break any new ore or to proceed with any development work or any new operations in or about such mine" and also prohibited owners from removing "any ore or waste from such mine, either above or below ground, or [from] conduct[ing] any other operations in or about such mine, except to the minimum amount necessary to maintain its buildings, machinery, and equipment in repair, and its access and development workings safe and accessible." (Taft Law 2016)

Scribner transferred from the Kenton to the Knoxville Mine.[1] Here he was returning to his family's roots. His grandfather was brought to Cali-

1. The War Production Board was a government entity created in 1941 to help convert civilian to wartime production. Its Limitation Order L-208 stopped the mining of all nonessential metals and virtually shuttered gold mining.

fornia as an infant, and in 1875 was a teamster carrying freight to Knoxville from Napa. Seeing opportunity there, he opened a hotel in Knoxville and began to acquire ranchland in the nearby Berryessa Valley.

This valley was submerged when the Bureau of Reclamation built Monticello Dam on Putah Creek in 1957. The purposes stated then included flood control, municipal and industrial water supply, and irrigation water supply. Pacific Gas and Electric Company began construction of the Monticello Power Plant in September 1981, and the first electrical energy was generated on March 9, 1983. The Lake Berryessa Recreation area is now administered by the U.S. Department of the Interior, Bureau of Reclamation, Mid-Pacific Region. Families, including the Scribners, had owned land in the valley for generations, tending their crops and orchards and burying loved ones in the cemetery there. The grief and bitterness at the loss of that valley were still strongly felt and expressed years later. Peter Scribner in 1999 tells us,

> There's an annual reunion at Spanish Flat of people that were out at Berryessa, on Memorial Day, when people go to visit the cemeteries, just kind of an impromptu thing because of the cemetery being there. Everybody was buried in Monticello. (Scribner interview, KMD vol. 7, 434)

In 1942 Thornton Scribner moved with his young family to be general manager and superintendent of the mine where his grandfather had opened the hotel more than half a century earlier. In 1999 Thornton's son, Peter Scribner, recalled how he helped in the war effort as a young boy at Knoxville:

> While the miners were mining cinnabar, the kids were "mining" scrap. We foraged around the dumps for... busted mine car wheels and axles, old engine parts, old motors, mine rail, pumps.... Periodically all this junk would be loaded into the mine dump truck and taken to a collection center in the little town of Winters, where it was weighed. (Scribner interview, KMD vol. 7, 431)

The superintendent of the Knoxville Mine had a well-built three-story house, rent-free. When the Scribners arrived in the 1940s, there was no electric power in the house. Kerosene fueled the stove and kerosene lamps and candles furnished light. For refrigeration, there was a canvas cooler that cooled by evaporation of water dripped onto it. A hillside spring supplied good fresh water, a welcome amenity (Scribner interview, KMD vol. 7, 425–26).

A signature sound of a mine operation was the whine of the drill sharpener in the blacksmith shop. A blacksmith course was required for an engineering degree when Herbert Hoover, the thirty-first president of the United States, graduated in Stanford University's first class in 1895. Legend has it that he met his future wife, Lou Henry, in the blacksmith class where she, a freshman, wanted to learn to shoe her own horses. She majored in geology and they were married after she graduated. He then became a successful mining engineer. Although his name is listed first, she is probably largely responsible for translating the classic treatise on geology, *De Re Metallica*, by Georgius Agricola (the pen name of Georg Bauer), written in Latin in 1556. The English translation, by Herbert Clark Hoover and Lou Henry Hoover, was published in 1912 by *The Mining Magazine*, in London. It was a significant contribution to mining literature (White House n.d.).

In the time that Peter Scribner recalled, a miner picked up sharpened drill rods at the beginning of his work shift and left them at the end of the shift. The ends—the bits—dulled from biting into hard rock, were then sharpened at the blacksmith shop for the next shift. Detachable tungsten-carbide drill bits, developed in Sweden after World War II, did not need daily sharpening and the blacksmith shop was outmoded. Scribner recalled the earlier routine:

> They were still using the old iron [drill] bits. They hadn't gotten the carbide insert bits yet.... They sharpened them. They had a whole blacksmith shop...which was a paradise when you're a kid.... There was no place that was off limits... [except] we could not go down the shaft. (Scribner interview, KMD vol. 7, 438–39)

The ore went from the mine to a jaw crusher, and then to the smelter, or furnace, as it was also called, that ran twenty-four hours a day. The mercury fumes were condensed and concentrated on a sloping table. The freed mercury ran out at the bottom, was put in flasks that were trucked to the Selby smelter, and then shipped out on river boats—the *Delta King* or the *Delta Queen* (Scribner interview, KMD vol. 7, 439–42).

As mining became more professionalized in the late nineteenth century, maps and mine models were used. One important responsibility of a mine manager, until the time of computers, was to maintain the mine model, a three-dimensional record of the mine operations. The model at Knoxville was a rack holding layers of glass sheets that represented the levels in the mine and were marked with colored ink to show the progress of the work (Scribner interview, KMD vol. 7, 448).

Mining historian Eric C. Nystrom writes that this type of model was first developed by a civil engineer, Charles T. Healey, for the New Almaden mercury mine near San Jose, California, and exhibited in 1874 at an annual trade show sponsored by the Mechanics' Institute in San Francisco. The model consisted of twenty-six glass sheets, each approximately twenty-six inches long and eleven inches wide, set vertically in a notched frame that held the plates exactly an inch apart. Each glass plate represented a vertical slice of the underground geology. The editor of the *Mining and Scientific Press* reported,

> It is not only an exquisite work of art, but it constitutes the most palpable, truthful and comprehensive work of representing the underground workings of a mine which has ever been devised. No description can do it justice. (Nystrom, *Seeing Underground*, 129)

The model that Scribner maintained in the 1940s at Knoxville was substantially the same, except it was horizontal. Later, quite different three-dimensional models would play an important role in the development of a modern gold mine next door.

Modern equipment replaced the pick and shovel. Column-mounted rotary rock drills ran on compressed air; diesel generators manufactured by Caterpillar supplied the power. Established companies like Joy and Ingersoll-Rand that had supplied mines for nearly a century were turning out up-to-date earth-moving and ore-processing machinery. At the Knoxville Mine, the innovative EIMCO rocker-arm loaders began open-pit operation next to the old shaft.[2]

Morgan North was one of the last mercury mine operators at Knoxville, in the early 1970s. He was founder of Howell-North Press and nephew of Julia Morgan, noted California architect who had designed the Hearst Castle at San Simeon, California. James Jonas delivered fuel to Knoxville and remembered Morgan North.

2. Western Mining in the Twentieth Century Oral History Series, Joseph Rosenblatt, "EIMCO, Pioneer in Underground Mining Machinery and Process Equipment, 1926–1963," an oral history conducted in 1991 by Eleanor Swent, Regional Oral History Office, The Bancroft Library, University of California, Berkeley, 1992.

> I was kind of amazed when I...met him the first time sitting on a huge excavator and he got off and talked to me for a while, and he sounded more like a university professor than a typical mining type of fellow. He was, as you know, involved in publishing and whatnot in Berkeley. He impressed me like this is his hobby, or he just really loved mining. The trouble was that was in an era when the environmental, and the pricing of ore, and everything was coming to a point where it was just impossible to make it with quicksilver.... I can remember Morgan telling me that the Bay Area Pollution Control District was coming down on him pretty hard about emissions. (Jonas interview, KMD vol. 4, 205–6)

In 1965 James William Wilder was attracted by the rising price of mercury and determined to try just once to succeed as a miner; he called his enterprise the One Shot Mining Company and leased the adjacent Manhattan Mine from the Knox family. He also enjoyed being with Morgan North, and they often got together at Wilder's place.

> Morgan was a pretty good friend of mine. He was running Knoxville for the Gambles, Launce and George. It was Morgan North Mine Management on a contract with Gambles. He was a pretty good technical person. He liked work. He could work like a dog even on a shovel. And this is so crazy, people don't understand that. This is a guy that owns a publishing company, and a pretty good publishing company. And what is he doing up here running a shovel? Well, that's what he wanted to do. I can still picture him: he'd be chewing on a cigar; he never smoked the cigar, he chewed the cigar.
>
> Morgan did find up in the attic of the old bunkhouse down there, some boxes up in the attic, and all these old bills of lading, shipment from Knoxville Mine to so-and-so, and it would be like 25Ks, 15Ms. And then the next week it would be maybe 25Ks and 25Ms. And he said, "I'm positive I know what this is. K is Knoxville, M is Manhattan. They shipped together." (Wilder 1996, 110–13)

Dean Enderlin, geologist at the McLaughlin Mine and a Napa County native and historian, knew of healthy and active old men who had worked in the World War II mercury mining boom when fulminate of mercury was still used in rifle primers and detonators. There were other mercury mine workers, from the more recent boom of the 1960s caused by the development of mercury batteries; two decades later, these workers were still healthy.

Dean Enderlin's grandfather, Henry Enderlin, was born in Baden-Baden, Germany, in 1882. In 1884 the family came to California, and settled in Lower Lake in 1896. Henry, then fourteen years old, went to work in quicksilver mines, and in 1905 was at the Knoxville Mine. After his wife died, he raised five children alone, baked five loaves of bread daily, and lived to be 101.

Henry's son, Elmer Enderlin, born in 1912, during his long lifetime worked at fifty-eight mines, including several mercury mines. It took some time to arrange an interview in 1995 because in the summer he worked his tungsten prospect in Idaho, and in the fall he was busy hunting elk. He came back to Lower Lake for the winter and then he recalled his time at the Knoxville Mine. His father had worked there earlier, in 1905.

> He told me that that mine was pretty well worked out at that time. In the heyday of that mine it was further back.... I was twenty-one when I started working at the mine. I had a ranch out there.... I went to the Knoxville in '40, I believe. (E. Enderlin interview, KMD vol. 3, 75, 77, 81)

Like Peter Scribner, Enderlin recalled the steel drill rods that had to be sharpened after every work shift.

> [They] had old blacksmith steel. They would put them in a forge and heat them and temper them and everything. I would have to take them out every night up to the shop, and then take the sharp ones down.... They built us all new houses down there when I was there. We were just moving into them when they bombed Pearl Harbor. (E. Enderlin interview, KMD vol. 3, 82, 86)

Elmer Enderlin worked sinking a shaft on a fifty-five-degree incline, and then driving a horizontal drift. He never worked around the retort and was aware of the risks there. He quit working at mines when he thought it was risky.

> That's one reason I've lived as old as I am. I would quit whenever it didn't look good to me; I would quit and go someplace else. I quit the Sunshine mine [in Idaho] the first shift. Several of them I quit the first shift. (E. Enderlin interview, KMD vol. 3, 91)

Enderlin enjoyed writing poems and won a national contest for poems not exceeding twenty lines. His was only sixteen lines and was published in *Treasured Poems of America* (Enderlin 1995, 425).

> ***The Prospectors***
> *Just a-sittin' on a log*
> *A-pettin' my dog*
> *Waitin' for snow*
> *So south we can go*
> *And leave these mountains sleep*
> *Under the snow so deep.*
> *We'll rest up until June*
> *Which will be all too soon*
> *And come back in the spring*
> *And do the same thing.*
> *We'll dig some more holes*
> *Like the lowly moles*
> *And open up the vault*
> *And if it's empty, it's not our fault.*
> *We'll quit that first million*
> *And start on the billion.*
> (E. Enderlin interview, KMD vol. 3, 109)

William Kritikos was born in 1922 in Tulsa, Oklahoma, graduated in civil engineering from the University of Oklahoma, and served in World War II as a combat engineer. He saw a better future in mining than in civil engineering,

> particularly for an entrepreneurial type like myself.... The professions are not all that different. If you forget about the metallurgy part, and a little bit of geology, there's a lot of similarity between the two professions. (Kritikos interview, KMD vol. 5, 1–2)

Following a successful mining partnership venture in Oregon, Kritikos came to the San Francisco Bay area in 1948.

> I said, "I'm going to get a job as an engineer, going to provide a good living for the family, but I'm going to do my romancing and mining on the weekends. It won't interfere with taking care of the family." So that's what I did. (Kritikos interview, KMD vol. 5, 5)

He was hired as engineer in charge of construction for the second Mokelumne Aqueduct for the East Bay Utility District.

> I decided that mercury was the place to go.... I'll tell you why: because Uncle Sam had announced a program for supporting the price of mercury at $225 a flask, because they were using mercury at that time as a coolant in the nuclear engines on nuclear power plants on the subs. So Uncle [Sam], in effect, guaranteed mercury to small producers. (Kritikos interview, KMD vol. 5, 6)

In 1955 Kritikos leased the Oat Hill Mine; forty years later, he and his wife Jacqueline still lived there, giving their address as Rancho Cinnabar, near Middletown, California. He processed the waste dumps and reopened the mine. After diplomatic relations were established with China in the early 1970s, he sold cinnabar to that market, where it was used in medicines and for decorative lacquer.

> It was an easy ore to beneficiate because of two reasons: the cinnabar occurred in...sandstone.... [You] can crush the sandstone down to about an eighth of an inch. That's pretty coarse, see? And by crushing all your sandstone down to minus one-eighth, you free all the cinnabar. That's fact one: you can free it very easily. That's the way the Good Lord put it in the ground. Second thing is, you can also beneficiate it very easily. The same way that you can pan it, you can concentrate it on a table.
>
> ...[It] was a two-table plant: two concentrating tables, a trommel screen, a nozzle that washed the material into a launder over to the screen. Then the screen material would go to the tables, and it would be concentrated. And then we would accumulate the concentrate in 5-gallon buckets. If you had a 5-gallon bucket about full of the concentrate, that would yield a flask of mercury, approximately. (Kritikos interview, KMD vol. 5, 17–18)

He knew that there were risks, and you had to be careful:

> My partner, Ward, was retorting one day, and he got careless. Now one of the things that you know when you're running a retort is that you have a double door seal and you have to carefully seal it off. You have to put the first door in, seal it with fire-ash and some other stuff; then you put the second one in, and when you do all this, you wear a mask, because that's the only dangerous time in the whole thing. Well, he didn't do it, and he had a bad seal, and then he did the foolish thing...instead of just shutting the fire off and walking away, which is what he should have done, he tried to correct it while it was

still heated up. And so he...started having trouble with his teeth, went to the doctor. He had gotten—what we term it is, got "squiffed."...meaning, he ingested some mercury vapor.

Now, some facts: that's not the end of the world, because a lot of these miners in the last two thousand years in the Almaden mine had been getting squiffed from time to time.... There's a lot of old-timers up here in these mines, here in California and elsewhere, that I'm sure had the same experience that my partner did...yes, you could get squiffed...your teeth might fall out, [you] start having aches and pains.... Do you die? No.

Now, you can take a teacup full of mercury and drink it yourself. A, you would not die; B, it would go through your system. I know you probably don't believe this, but it happens to be factual. (Kritikos interview, KMD vol. 5, 45–46)

Kritikos and Jackie visited the ancient Almaden mercury mine in Spain:

[In] this little town here you have old guys ninety years old, you have little babies toddling along, right in this town, and this town has been sitting on a mercury mine for two thousand years, and that's what these people all get their living from! So the question that flashed through my mind is how can we have this demographic profile here? Why isn't everybody dead?...[The] Belgian mine manager took us in, rolled out the red carpet, took us around the plant and the mine. Now, he had just built a brand-new plant; American design. Even though it was brand new, he had leaking pipes and everything. Here was mercury dripping out of these pipes down to pans on the floor. So he said, "Well, what do you think, Bill?"

I said, "Well, it's a good design. It's American design. But if you ran this thing up in California, you'd have your ass shut down at six in the morning. You couldn't get away with this."

"Oh," he says, "You Americans worry a lot about that stuff. Why do you think we've been able to operate this mine for two thousand years?" (Kritikos interview, KMD vol. 5, 42)

John Livermore, who called himself a prospector, and who was still active until he died at ninety-four, recalled visiting the Oat Hill Mine when he was young.

Oh, yes, we used to go out and visit when it was operating during the war. I remember going out there and they had a big vat of mercury

> and we would stick our hands in and just play with that mercury.... [Now], everybody says it's so dangerous that you can't even touch mercury; that's kind of ridiculous, of course. (Livermore 2000, 6)

Hugh Ingle was born in Santa Rosa, California, in 1923. His father attended the Michigan School of Mines and operated small pocket gold mines in California and Oregon. From the time Hugh was ten years old, he helped in the mines. He graduated from the Mackay School of Mines in Reno, and was a fighter pilot in the Korean War. He was always an owner and/or operator, and an advocate for small mines. He worked at the Oat Hill and Reed Mines, and in 1955 he leased the Corona Mine from John Livermore and reopened it. Livermore remembers the operation:

> Hugh and his son Hughie constructed almost singlehandedly a plant for roasting the ore that consisted of a crusher, rotary kiln, and condensing system. This fairly simple recovery method resulted in the production of pure mercury by condensing the volatized gas produced by heating the ore. They also constructed living facilities adjacent to the furnace. This is ironic when one considers the overstated fears of the dangers of mercury in our modern economy. The metallic or inorganic mercury is not the dangerous form. It is only when it is converted to organic mercury in streams that it enters the food chain and is a health problem. They must have been breathing a certain amount of the gas, but it didn't seem to affect their health. (Ingle 2000, x)

Ingle recalls that he and his son often slept by the furnace, to keep warm and to maintain the twenty-four-hour operation.

> The ore was pretty tough to handle, you had to retort the ore about two or three times to get the mercury to come out. This retort held 500 to 1,000 pounds of ore. Twenty-four hours a day I was watching it, about every four [hours] I'd go out and check it. Sometimes when we were having trouble with mud or something like that plugging up the feeder, I would sit up on the feeder and half the time go to sleep. And, you know, when we slept up there on the dump, and even when we were in the cabin here, if there was a change in the noise that the furnace made, you knew it and you woke up right away. But I would sit on that feeder because it was warm up there. It was handy because you had to poke the mud through if it plugged up. But it was a real good operation. (Ingle 2000, 8, 9, 26)

It became unprofitable in 1970 so he shut it down, and then returned in his late seventies, as contractor for reclamation.

Anthony (Tony) Cerar was born in 1913, and grew up on what became the Livermore property, near the Oat Hill and Corona Mines. The Twin Peak Mine adjoined the property. He was twenty-two when he went to work at La Joya Mine, in 1935. During his lifetime, he worked at many mercury mines in or near the Knoxville area: Abbott, Baker, Bella Oak, Dewey, Granada, Helen, La Joya, La Libertad, Mirabel, Oat Hill, Palisades, Rinconada. For a very short time, he worked on the furnace at the Klau Mine in San Luis Obispo County; he left because the furnace leaked and workers were getting sick. At the Abbott Mine he sometimes worked fifteen hours a day at the furnace. At other mines, he worked at the full spectrum of jobs, from laborer to part owner, which in most cases were the same. Cerar recalled the situation at the Redington/Knoxville Mine in earlier times.

> The second time that Knoxville was running in a big way was when the Boston Company took over in 1900.—Jake [Jacob Carl] Ansel was an uncle of mine by marriage and was a foreman at the Oat Hill, and they sent him over to Knoxville to take charge of reconditioning the Redington shaft that had been unoccupied or unused for about eighteen years or something like that. After they got everything going, the furnace going, they developed a lot of ore there to start with. A lot of these people were coming over from Europe and took the place of the Chinese. They worked for the same wages. My aunt told about a lot of those Austrians or Yugoslavs that came over to Knoxville; they put them on the furnace right away, and they came over here and had been living on vegetables more or less in the old country—pink cheeks and everything else, healthy looking, and after they had been on the furnace a few months and lost their teeth and sallow-looking and all that. See...most of the managers of the mines didn't care at all. (Cerar interview, KMD vol. 2, 27–28)

Tony's father worked at the Knoxville Mine at the beginning of the twentieth century, and in 1945 Tony and a partner leased the mine from George Gamble. Gamble, always generous, rented equipment to them now at such a low price that it seemed to Tony it was given to him. In 1995 Cerar recalled it as "the best deal I ever had":

> We had a couple of men working for us there...we had a cook...and we lived in the schoolhouse. We started out with some dump material...on ore that the old-timers had left behind.... The Caterpillar dozer dug most of that rock...we just continued working in the open pit there after the price of mercury came up again then went down. The price beat us out too, eventually.... It started out at around $150 a flask and went down to way below eighty at the last. (Cerar interview, KMD vol. 2, 40–41)

When he was in his eighties, he still climbed around the mine site with agility.

James William Wilder, the last owner of the Manhattan mercury mine, named his company the One Shot Mining Company. He knew that folks criticized his unorthodox setup, and spread rumors that mercury was "hanging off the trees" there. He also thought they exaggerated the dangers of mercury.

> I tell them, "Hey, wait a minute. Let's get one thing straight. If the guy is running a mercury operation, what is he getting paid for?" "Well, he's getting paid for mercury." "Well, if he's got it dripping off the trees, the guy obviously is not getting it in the flask and I don't know of any miner that's that stupid." It's like having a gold mine and saying, "He polluted the creek. It's full of nuggets down there." No, that's baloney.
>
> I had a guy come out here and looked at the trees and said, "Ha. I see you've killed all the trees. The leaves are all falling off."
>
> I said, "No, it's fall." Jeez, I think fall means the leaves fall off, don't it?
>
> Then he came back in the spring and he looked at his notes and he said, "Well, I can't believe it. I thought the trees were all dead. They've all got leaves on them."
>
> I said, "No, they do that every year. The leaves fall off and then they get them back on." He said, "I'll be darned."
>
> We monitored ourselves all the time. Every week, I got some urine bottles back to the labs back there and they'd run them on everybody, myself included. I'd keep score and we had a regular

> chart of everybody, where they were at all times. If somebody started getting a little bit elevated, moved him outside and moved somebody outside inside because in hot weather, if you're working in a place where there's spills or bottling mercury, well, you do have some fumes even with a respirator. (Wilder 1996, 97–98, 99)

Bill Wilder's operation was going to change, from mining mercury to mining something quite different after 1978.

CHAPTER 3

LAKE COUNTY, CALIFORNIA

Someday they're gonna find some use for this worthless country.

—Della Conner Underwood, Knoxville rancher

Clear Lake, about seventy miles north of San Francisco, is the largest freshwater lake that is entirely in California; according to some geologists, it is the oldest lake in America. It lies at the base of an extinct volcano, 4,200 feet in elevation, that is called Konocti by the Pomo, Miwok, and Wappo people who first lived there. According to the US Geological Survey (USGS), Mount Konocti is around 300,000 years old, and the most recent eruption in the Clear Lake volcanic field occurred about 11,000 years ago.

The American settlers who arrived in the 1840s were farmers who planted orchards in the rich soil of decomposed lava. They named the lake Clear Lake, and in 1861 the governor of California approved Lake County, with the town of Lakeport as county seat. It is the only one of California's fifty-eight counties that never had a railroad.

In 1872 residents of Napa County, with a larger population, coveted the water from the lake, as well as the profits from the Knoxville mercury mines to the southeast. They paid Lake County $3,500 to acquire the Knoxville Township, separated from the Napa Valley by mountain escarpments. The March 8, 1872, Act of the Legislature "provided further, the Board of Supervisors of Napa County should order paid the claim of Lake County for the sum of $3500, and the Auditor of said county of Napa should draw a warrant for the sum on the Treasurer of the said county, payable from the general fund, and that the Treasurer of Napa County should pay the same" (Lyman 1881, 112).

The disruption that caused the volcano, and the steam fields around it, also created a defining geologic feature of western California: the fault that extends 810 miles from Eureka, in the north, to Brawley, near the Mexican border. Here the North American and Pacific tectonic plates slip by each other, moving about two and a half inches a year. The fault was identified in 1895 by Andrew Cowper Lawson, a professor of geology at the University of California. Lawson was known for his egotism, but he did not name it the Lawson Fault; instead he chose the Spanish equivalent of his first name, and not-so-modestly named it the San Andreas (Saint Andrew) Fault. This fault ruptured on April 18, 1906, and San Francisco was severely damaged by the subsequent fire. The State of California then sponsored the committee, chaired by Lawson, that produced the two-volume report on the earthquake, a classic in the field of seismology that became known as the Lawson Report (UC Berkeley n.d.). Lawson's (1908) report remains the authoritative work, as well as arguably the most important study of a single earthquake.

Gordon Oakeshott, deputy director of the California Division of Mines and Geology from 1948 to 1974, was Lawson's student in 1924, and recalled,

> The great Andrew C. Lawson was an old curmudgeon, but we liked him because he was so enthusiastic. Lawson wouldn't brook any disagreement with him, but I got along fine with him because he knew his stuff and he was an inspirational speaker. You know, he was the old professor who became notorious for having a child at age eighty-seven.
>
> After the Lawson report had been brought out and digested, the reaction in San Francisco and the Bay Area and California was that we should keep the matter quiet. In 1906, the time was not ripe for people to take cognizance of earthquakes. "We don't want California to get the reputation of being dangerous because of earthquakes." They said the so-called earthquake of 1906 was mainly a fire. But of course the fire was due entirely to the earthquake. Water supply was cut off, gas mains were broken. (Oakeshott 1988, 8, 13)

Geothermal activity still influences much of the region. The misnamed geysers—they are technically fumaroles—in the area were once a world-famous tourist attraction. Visitors to the Geysers Resort Hotel included Ulysses S. Grant, Theodore Roosevelt, and Mark Twain. Today the Geysers steam field, the world's largest geothermal field, with more than 350 steam wells deep underground, produces enough electricity for about two cities the size of San Francisco.

At the head of Napa Valley, known now for its wineries, settlers in the nineteenth century found hot springs that reminded them of the spas of Saratoga, New York. They coined the name Calistoga, and tourists still come there for healing and enjoyment. Rockhounds (collectors) and amateur jewelers come to hunt for the unusual yellow opals to be found in the sinter around hot springs and geysers.

In the years after World War II, Clear Lake became known as the "poor man's Tahoe." The Plumbers and Pipefitters Union in Northern California, Local 38, run by Joe Mazzola, built the Konocti Harbor Resort in 1959 as a vacation spot for its members, and also, it was rumored, as a place for carrying on shady dealings. It became the largest concert venue on the North Coast. Fishermen caught bass, despite concerns about mercury seeping into the lake from the Sulphur Bank Mine at its edge.

California State Route 29 takes off from Interstate 80 just north of the Carquinez Bridge, where the Sacramento River drains into San Pablo Bay, the northern extension of San Francisco Bay. At first a four-lane expressway, Route 29 goes north and west through the cities of Vallejo and Napa, and then becomes a two-lane road past vineyards and wineries. At the head of the Napa Valley in Calistoga, it climbs north over the flank of Mount Saint Helena, another extinct volcano, elevation 4,344 feet. It enters Lake County, goes down to Middletown, and over another hill before reaching the town of Lower Lake.

Here there is a four-way intersection. Highway 29 turns to the left, to wind north and west around Clear Lake, past Konocti Harbor, skirting Mount Konocti, through the town of Lakeport, and on to Upper Lake.

If one turns to the right at the intersection, Main Street goes a short distance through a business district to the edge of town, past the cemetery, and then becomes State Route 140, the Morgan Valley Road, going on to Knoxville, now abandoned but once known for its rich mercury mines. The road from there on is marked impassable or wilderness area on modern maps. At the end of Morgan Valley, the road climbs up to a watershed, and the junction of three counties: Lake, Napa, and Yolo. Along the way, the road hugs the side of a mountain, and curves to give a stunning view of low mountain ranges, chaparral and oaks, and an occasional gleam of pond or waterway. No towns are visible, and the orchards of Capay Valley, the vineyards of Napa, and the mines of Knoxville are all hidden by mountains. It is easy to think that this is a land where no one lives or ever has lived.

The local Audubon Society actively promotes protection of bird habitat bordering the lake. One of its members, Roberta Hanchett Lyons, recalls:

> I liked Morgan Valley...driving out on that road and thinking, "This is so neat." There were no power lines. It was really pristine in that

> it was untouched. You could look out across the vast hills, and there was no evidence of human beings. (Lyons interview, KMD vol. 5, 113–14)

She doesn't notice, on the far skyline, the rusting headframe of a long-gone mercury mine.

EVERYBODY WAS RELATED OUT THERE THEN

Jack Landman's grandfather arrived in Morgan Valley in the 1800s, and Jack worked the family ranch there until 1972. The ranch was on Reiff Road, named for his step-grandfather. Jack, born in 1915, attended the Morgan Valley school. He went to live with an aunt in Berkeley, graduated from high school there in 1932, and returned to the ranch. He said,

> Everybody was related out there then.... When I started ranching, I didn't know much about ranching, so my neighbors helped me out. We farmed with horses at that time. We did that until we bought our first tractor and that was in the late thirties.... It was what they called a Fordson. It's the first tractor that Henry Ford put out.
>
> [We] had sheep, too. And then we sold the sheep and went entirely into cattle.... I started buying calves and raising them up and selling the ones that I wanted to sell and keeping the breeding stock, and that's how we built up our herd of cattle. (Landman interview, KMD vol. 5, 62–63)

During World War II he worked as a laborer on the surface at the Reed mercury mine, for which he was paid $150 a month. He praised the Bradley family, owners of the mine:

> See, you got your meals...three times a day. So that went along with the job...The Bradleys fed really good.... You had salads and soup and meat. Good, solid foods. Ice cream and cake the cooks would bake. (Landman interview, KMD vol. 5, 69–70)

He was never bothered by cattle thieves, but the deer poachers were a nuisance and a growing threat.

> [It] was hunters all the time, spotlighters...that were shooting around. They were opening roads in behind us, and they would go in behind and hunt. You'd run into some pretty rough characters sometimes back in the hills. You wanted to be sure of having some

> kind of protection.... Even at night I packed a gun sometimes, but never got into a shooting scrape.... [It] was getting worse and worse. (Landman interview, KMD vol. 5, 81)

He sold most of his ranch in 1972 to a group who wanted to develop homesites.

IT WAS VERY TAME

Bonny and Ross Hanchett both had degrees in journalism from Washington State University, and bought the *Clear Lake Observer*, a weekly newspaper, in 1954. Bonny recalled Lake County as it was then:

> It was very tame. I mean, we had no problem with crime or with gangs or with anything. It was really nice. I could let my children go over to the Highlands and not have to worry. They had the miniature golf and the swimming beaches.... But I had no fear. People would wander up and down Lake Shore Drive in the early evening, and it would be very pleasant. (Hanchett interview, KMD vol. 4, 77)

A FUNNY, OLD-FASHIONED STORE, THE HUB OF THE COMMUNITY

In 1971 Beverly Magoon and her husband lived in Walnut Creek, California. She was a weaver, active in the artist/craftsman community in nearby Berkeley, and he was a research chemist reluctantly facing a transfer to Texas. They had three teenage boys and a younger daughter and wished they could find a quieter life for them. He always liked hardware stores, and one evening he saw in the newspaper an ad that caught his eye: an old-fashioned general store in Lake County was for sale. She recalled,

> And so when we came up here and we drove into town, I said, "Ah, I like this. I like this town." We parked under some trees across the street from the old store, and I said, "I think we're home. This is the place." That's how we started up here. (Magoon interview, KMD vol. 5, 206)

They bought the store on Main Street and sold groceries, hardware, and fabric; for a while, they even had a deli and butcher shop. Their store was the only one in the area that sold feed:

> We had all kinds of feed. We had chicken feed, horse, pigeon, wild birds, dogs, cats, just about anything that needed food, we carried it.

> And we had a butcher shop for a while.... We had feed, and we had groceries, and hardware. And we had horseshoes.... [We] just had a little fabric. We didn't have a lot of fabric. We had Aladdin lamps. And I remember the first Aladdin lamp I sold, and I thought it was so expensive, was like $16.95. They're now, well, I have one up there that is $249, so you can see how that has changed in twenty-five years.
>
> It was a funny, old-fashioned store. Really, at that time, kind of the hub of the community.... And eventually we would have parties once a year in back of the old store. We'd get a band and we'd have kind of a hoe-down. Everybody sat on hay bales and brought potluck. Potluck would consist of something like a whole turkey. People brought things like that.... [We'd] have fifty, a hundred people, which was quite a few people back then. (Magoon interview, KMD vol. 5, 209–10)

YOU USED TO LEAVE YOUR HOUSE UNLOCKED AND THE KEYS IN THE CAR

James Jonas, born in Santa Rosa in 1938, was from another pioneer family. His grandfather sailed around the Horn from England to California sometime in the 1800s, and worked as a teamster in Occidental, a small town west of Lake County. Since then, the family always had their Thanksgiving dinner at the Union Hotel in Occidental. Jim's father ran the bulk plant for Union Oil Company in Lower Lake. Jim attended Lower Lake High School and played trombone in a dance band for the high school dances. When he graduated in 1955 there were only 125 students in the high school, and young people made their own fun, as he recalled:

> [You'd] visit one another. I don't know, we always seemed busy.... [There] was a litle bit of beer every once in a while; no drugs I... heard of.... There wasn't any thievery, or vandalism, or things like that. I'd say that probably the raciest thing we would do was haul a car body down the street upside down on Halloween.... You used to leave your house unlocked and the keys in the car, and even if you went away for two weeks, just in case somebody needed something. (Jonas interview, KMD vol. 4, 86–87)

> So when you had the summer influx, it was really noticeable. I mean you went from, you know, in the middle of the winter you could

shoot a cannon off down the street and not hit anybody. In the summer, there were people walking, and particularly in a town like Clearlake Highlands or Lakeport, people strolling the boulevard, a lot of people around. Tourism was a much bigger part of the action.

...And of course, agriculture...grapes weren't much in those days, but pears and walnuts was a big part of the economy here. So I'd say agriculture and tourism was about it. (Jonas interview, KMD vol. 4, 193–94)

Jim Jonas attended the University of California, Berkeley (UC Berkeley) and worked at Bank of America as a management trainee in Oakland. In 1960 he married a high school classmate, and when they realized they were homesick for Lake County, they moved back there. At first, he managed a beer distributorship, with a territory including Oregon, Washington, Idaho, Montana, and Nevada. In 1970 his father had a stroke, so Jim took over the bulk plant in Lower Lake.

[When] I came up here in 1970, he only had one delivery truck, which was a 1948 Chevrolet that held 1,150 gallons.... He probably had 60,000-gallon storage.... So I came up here and took over this little rinky-dink plant and his 1948 gas truck, and I was just as happy as a clam to do that, which led into dealing with the likes of Bill Wilder and Morgan North in the quicksilver mines. (Jonas interview, KMD vol. 4, 193–95)

Jim supplied fuel, mostly diesel, for some of the construction work of power plants at the Geysers, and also to smaller businesses:

[Typically], stations even in those days, say in the Bay Area, would be directly supplied by the oil company, but in the rural areas where the stations were smaller and the storage was smaller, they wouldn't deliver them direct from, say, Richmond up here because they'd have to take 8,000 gallons at a time and full truck and trailer. Most of these little stations didn't have that kind of storage, so we took care of them.

[Our] clientele in those days were farmers, you know, that have a 500-gallon tank out in their farm yard somewhere, or these contractors that would be up on a job with an old fuel truck or we would put tanks out there for them, deliver it directly into their construction equipment. So we had that type of business. (Jonas interview, KMD vol. 4, 196)

Bill Wilder's One Shot mercury mine was about twenty miles away on Morgan Valley Road, and Jonas furnished fuel there.

> In those days it was a dirt trail; I mean, it would take you half a day to get out there, almost, particularly in the wintertime. (Jonas interview, KMD vol. 4, 200)

NEVER EVEN GAVE A THOUGHT ABOUT THE GOLD

Della Conner Underwood's parents were ranchers and living in Knoxville when she was born. She attended the Knoxville schools, graduated from Napa High School, and married Curt Underwood. In 1970 she and her husband moved back to the Knoxville house that her parents lived in when she was born. They continued ranching in the area, and leased land from Bill Wilder at the Manhattan Mine location for grazing their cattle. She recalls,

> We got a pretty good deal from Bill [Wilder]. Curt gave him a steer a year for the rent.... You know, it's really funny. One time I can remember we were getting some cattle out of the Manhattan, and we rode up there horseback, and I can remember riding up there, and just about where the truck shop is down there now and kind of looking off out toward Hunting Creek, you know, at all the sagebrush and stuff and all the old tarweed growing around, and I said to my husband, "Someday they're gonna find some use for this worthless country around here." I said, "They'll make medicine out of the chamise brush or something." Never even gave a thought about the gold. (Underwood interview, KMD vol. 7, 334)

In 1978 Knoxville township still had few telephones, surfaced roads, or bridges. Napa historian Robert McKenzie called it "truly the last frontier of Napa County." This was about to change, due to events occurring far from there.

PART 2

CHAPTER 4

HOMESTAKE MINING COMPANY

Our civilization depends on reconciling natural resource extraction with environmental concerns.

—William Langston, Homestake vice president and general counsel

In mid-October 1877, San Francisco mining investor George Hearst purchased a claim on the north fork of Gold Run Creek, near the present town of Lead, in the Black Hills of South Dakota. The claim had been staked by French Canadian brothers Fred and Moses Manuel, with the hope that it would pay enough for them to return home: that it was a home stake. Hearst went back to California, and, with partners, formed Homestake Mining Company, incorporated as one of the first California corporations on November 5, 1877. The five original directors and subscribers of capital stock were all residents of San Francisco, and each of them held five shares of capital stock valued at $500 each. The company was immediately listed on the San Francisco Stock and Mining Exchange, and in January 1879 it was the first mining company listed on the New York Stock Exchange, where it would become one of the longest-listed stocks in the history of that exchange.

In 1977 the Homestake Mining Company published a book by William Bronson and T. H. Watkins, *Homestake: The Centennial History of America's Greatest Gold Mine*.

> The Hearst legacy would remain unbroken right up to the mine's centennial year, first in the person of [Edward H.] Clark, who had been appointed to the board of directors by Mrs. Hearst in 1892, had become president in 1914, and would remain so until his own death in 1944. And now, in the person of [Donald H.] McLaughlin, a close

> personal friend and protégé of Mrs. Hearst's as a boy and young man, who joined the company in the summer of 1926 as a consulting geologist and would remain so until 1942 when he joined the board of directors. (Bronson and Watkins 1977, 62)

The history explains that "near the end of 1925 was a time of uncertainty" about the extent of the Homestake lode (Bronson and Watkins 1977, 61–64). McLaughlin had worked at Hearst's Cerro de Pasco Mine in Peru as a geologist, and "by 1925 had established such a reputation that he was offered a full professorship at Harvard in mining engineering—the youngest professor the university had ever appointed" (Bronson and Watkins 1977, 63). He was recruited to spend the summer of 1926 "in wide-ranging exploratory work, and by the end of the season was convinced that the Homestake Lode was not only not going to disappear, but that it dipped deeper and farther.... Equally important to the future of the mine was the system of selective mining he introduced... [his] hard work and imaginative competence had been instrumental" (Bronson and Watkins 1977, 65).

The suspicion, sometimes antagonism, between geologists with their theories and miners with their "nose for ore," often existed in mining operations. Clarence Kravig graduated from the University of Minnesota in 1929, went to Homestake as a geologist, working under Lawrence Wright, chief geologist, and later became mine superintendent and then assistant general manager, so he saw both sides. When Kravig arrived at Homestake, he soon became aware of the rift between the geologists and the miners.

Kravig was interviewed in 1993 and gave his recollections:

> In 1926 Donald H. McLaughlin was named consulting geologist. He was a professor at Harvard...and a protégé of Edward Clark, the president of Homestake. That kind of posed another problem around here. The mine department under Alec Ross did not like to recognize geologists, but they had to respect the presence of the consulting geologist, who was named by the president of Homestake. (Kravig 1995, 11)

> Lawrence Wright mapped the geology, but it was quite technical, and apparently he couldn't sell it to the mine department. They were detailed geologic maps put on tracing cloth. Tracing cloth goes to

> pieces pretty fast if it is handled much. They were lying in the mine office drawers, just as fresh and clean as the day they were made; nobody in the mine department ever looked at them.
>
> In my association I had to go around to the various foremen of the mine and sometimes shift bosses. I was told by Mr. Albert Pendo, one of the foremen...that the mine department wouldn't take any cognizance of geology—wouldn't recognize the geologists whatsoever. They went their own way as hard-rock miners, and as some of them said, "I graduated from the school of hard knocks."...He said, "Oh, it would never do for me to... [admit] I had talked to a geologist, and that's taboo." (Kravig 1995, 7–8)

The miners and/or the geologists saved the mine by tracing the Homestake lode to further depths. In 1945 McLaughlin became president; in 1961, chairman of the board; and in 1969 honorary chairman of the board and chairman of the executive committee. The company history says, "Between them, [Edward H.] Clark and McLaughlin represented a continuation of the Hearst influence on the fortunes of the Homestake" (Bronson and Watkins 1977, 62). It praises "the almost visionary energy of the company's highest officers—particularly McLaughlin, John K. Gustafson, who became president in 1961, and Paul C. Henshaw, who took over the presidency from Gustafson in 1970... [both] had been students of McLaughlin's at Harvard. Like their former teacher, they were scientists" (Bronson and Watkins 1977, 71–72).

The centennial history does not mention the metallurgists at the Homestake Mine who were advancing methods to improve recovery of gold from the ore, and to contain and clean up mine waste. Allan J. Clark, no relation to Edward H. Clark, came to Homestake in 1897 as a Columbia University graduate; his longtime assistant and then successor, Nathaniel Herz, had degrees from Yale.

The centennial history tells how the company began its long nonunion tradition: In 1909 the World Federation of Miners/American Federation of Labor attempted to organize the workers at the mine in Lead. Before they could organize a strike, Superintendent Thomas J. Grier ordered a preemptive lockout, and on November 7, 1909, he announced that the Homestake Mining Company would employ only nonunion men after January 1, 1910.

The company successfully fended off subsequent union attempts by providing benefits for workers: free medical care at a well-staffed hospital, a recreation building that housed a free library, bowling alleys, and swimming pool. The adjacent theater charged a minimal fee. Good houses were built for workers and the company store offered high-quality clothing, furniture, and groceries. Operations always maintained a superior safety record. It remained a nonunion operation until 1966, when the United Steelworkers succeeded in winning an election there.

Throughout the Rocky Mountain mining belt, in Colorado, Idaho, and Utah, Homestake operations were nonunion. When the Ambrosia Lake uranium district boomed near Grants, New Mexico, Homestake personnel spearheaded the drive to build a first-class hospital, the only one in the area. They cooperated with major unionized companies, Phillips Petroleum and Kerr-McGee, to attract a developer, Jack Stagner, to build housing for thousands of workers. The houses were similar, except that only the Homestake houses had a doorbell and two bathrooms. Once again, Homestake fended off unions by providing extra benefits and maintaining a superior safety record. It remained the only nonunion company in the Ambrosia Lake District. Later, in California, this tradition would continue.

In 1975, even as the *Centennial History* was going to press, the company was changing both in management and in philosophy. It had extended exploration activities and operations worldwide and to a variety of minerals: base metals, uranium, and an ill-fated venture into clay products. The official U.S. price of a troy ounce of gold, which had been fixed since 1934 at $35, was freed in 1972, and by 1975 was $160 and climbing. Innovations in earth-moving equipment made more deposits economically viable, in open-pit mines as opposed to underground operations. The executives decided to reorganize Homestake and return to its historic role as a gold mining company. They hired a management consultant to locate someone who would be in line to become chief executive when Henshaw retired in three or four years.

Harry Milton Conger filled the bill and was hired on November 1, 1975. Conger was born in Seattle, Washington, in 1930, got his Social Security card when he was twelve, and worked in local apple orchards in the summer. Two months after graduating from high school, he married Phyllis Shepherd. He attended the University of Washington for two years, studying engineering. His uncle had a coal mine at Snoqualmie Pass in the Washington Cascades. Conger recalls:

> It was probably my uncle's coal mine that first twigged me to this. Mining engineers got to do all kinds of engineering: electrical, mechanical, geologic—and then all of the mine infrastructure itself,

each of which involves engineering one way or another. You got to work with people and that just seemed so much better than sitting in an office somewhere. And then I got to go to Cominco's Sullivan Mine up in British Columbia and got to go underground. That really tilted the scale for me. From then on, that's what I wanted to do. (Conger 2001, 11, 30–31)

He transferred to the Colorado School of Mines in Golden, receiving two scholarships, one from the Woman's Auxiliary to the American Institute of Mining Engineers (WAAIME) and the other from the American Smelting and Refining Company (ASARCO). He joined the Reserve Officers' Training Corps (ROTC). Phyllis supported them by working as receptionist and secretary for the Coors brothers.

After he finished college in 1955, he worked for ASARCO at the Silver Bell open-pit copper mine in Arizona, first as an engineer and surveyor, and then as a shift boss. He took leaves of absence to fulfill his six-month military commitment. In 1964 he went to Kaiser Steel Company's Eagle Mountain iron mine in Southern California, starting as foreman on the day shift and rising to mine manager. From there, he moved on to be manager of the Ballmer Coal Mine in Fernie, British Columbia, and then superintendent of mines for the Consolidation Coal Company in Illinois.

On November 1, 1975, Harry Conger became Homestake's vice president and general manager of the newly created base metals division that included a lead-zinc mine, mill, and smelter in Missouri, and a copper project in northern Michigan. In 1976 Conger took over the uranium division, with operations in Colorado, New Mexico, and Utah. In 1977 when Paul Henshaw became chairman and chief executive officer, Conger was made president and chief operating officer, the first mining engineer to run the company.

As Homestake was changing its management, public attitudes toward mining were changing. On January 1, 1970, President Richard Nixon signed into law the National Environmental Policy Act (NEPA). Later that same year, the California legislature passed, and Governor Reagan signed, the California Environmental Quality Act (CEQA). The first Earth Day, celebrated on April 22, 1970, sparked the concept of environmentalism, and resulted in a new profession—environmentalist. Senator Gaylord Nelson and Congressman Pete McCloskey sponsored legislation creating the Environmental Protection Agency and leading to passage of the Clean Air, Clean Water, and Endangered Species Acts.

Concern for the environment was not, however, entirely new to the staff at the mine at Lead. Since early in the twentieth century, they had worked for community health and protection of the environment, although they did not call it that. Early on, Homestake had provided drinking water from Spearfish Creek for the local communities, and carefully monitored it for purity. In time, the water was chlorinated to ensure its safety.

There had long been concern there for the wastewater flowing from the Homestake Mine processing plants into Whitewood Creek, partly because it might contain two valuable elements: gold and mercury. Research carried out by Charles Merrill in the 1890s led to additional treatment of the ore with cyanide following the amalgamation with mercury. New cyanide plants and slime plants were built, and recovery of gold rose from 75 percent to 94 percent. Merrill's process developed at the Homestake Mine was patented, adopted throughout the world, and brought him fame and wealth (https://www.mininghalloffame.org/hall-of-fame/charles-washington-merrill).

Amalgamation with mercury and the use of cyanide were gradually phased out, with the development in the 1970s of the carbon-in-pulp recovery process, and carbon adsorption. With these advances, gold recovery rose to 97 percent.

In the 1980s Terry Mudder's research at the Homestake Mine made further progress in environmental protection. On February 22, 2014, Terry Mudder was the first inductee into the International Mining Technology Hall of Fame in the environmental management category in recognition of his contribution to the industry. The citation read, in part, as follows:

> He is often considered the first to apply biotechnology in the mining industry. The development of the novel combined aerobic attached growth biological treatment process using rotating biological contactors at the Homestake Mine in Lead, South Dakota in the early 1980's was the first of its kind relying on bacteria to remove thiocyanate, cyanide, ammonia, and metals from process solution producing up to 15,000 cubic meters per day of treated wastewater that was discharged into a permanent trout fishery. ("Terry Mudder" 2013)

When uranium mines were developed in the 1950s in Colorado, New Mexico, and Utah, Homestake faced new challenges as radiation hazards were added to the usual concerns. Mine safety and workers' health were given high priority there, and the mine operations, with Langan Swent as manager, won national awards for their safety.

Although the mine operators had long been acutely aware of the importance of mine safety, workers' health, and environmental protection, these had not been primary concerns in the corporate headquarters. When Langan Swent was named vice president of engineering in 1975, his job outline had seven items, and the order probably represents the level of interest at the corporate level. The sixth item was to plan and coordinate the company's mine health and safety programs; the seventh and last was to monitor the political, social, and legal environment as it affects mine safety and health, and environmental issues.

Swent saw Conger's coming as a breath of fresh air:

> He accepted the need for concern for the environment.... [One] of the [important] things was to have the company make the transition into the modern environmental situation.... [When] a lot of this environmental business started, the reaction of everybody was: "Fight 'em, they're a bunch of wild-eyed radicals."... They [the environmentalists] weren't always 100 percent right, but it made more sense to work with them than try to work against them.... Harry has accepted that as a fact of life.... Harry also supported my efforts in health and safety greatly, the first person in Homestake at the presidential level really to show much interest in that. (L. Swent 1996, 867–70)

William Langston, Harvard law school graduate and Homestake vice president and general counsel from 1973 to 1992, also saw the importance of this issue. He recalled that George Simchuk, manager of a Homestake uranium project in Pitch, Colorado, had "absorbed values of the kind that are often referred to as 'touchy-feely,' values that I believe are necessary if natural resource extraction can be reconciled with environmental concerns. This is one of the great considerations of our time. Our civilization depends on it" (William H. Langston, personal conversation with Eleanor Swent, October 2015).

At the same time, other personal events happening offstage at Homestake were going to have consequences for the company as it developed a new mine in California at the junction of Napa, Lake, and Yolo Counties.

In 1948, five years after he became first dean of the College of Engineering at Berkeley and three years after he became president of Homestake, Donald McLaughlin met and married Sylvia Cranmer in Denver. She had graduated from Vassar with a major in French. Her family background was in mining and in what would come to be called environmentalism. Her father, George E. Cranmer, grew up in a ranching family and loved the outdoors. He attended Princeton, started a brokerage firm, was appointed Denver's manager of Parks and Improvements, and is remembered for promoting the construction of the Red Rocks Amphitheater.

Sylvia's uncle was William Henry Harrison Cranmer Jr., a major figure in the prosperous silver-lead mining district around Park City, Utah, before it became famous for its powder snow and film festival. He was president and general manager of New Park Mining Company, the leading producer in the Park City district (*Park Record*, May 5, 1943). In 1941 Union Pacific built a spur line there at a cost of $99,000. On Sunday September 21, nine hundred people attended the dedication ceremony, and a special silver spike was driven. This was to recall the memorable golden spike ceremony at Promontory Point, Utah, nearly a century before, when the transcontinental railroad was completed. The station at the end of this new line was named Cranmer (UtahRails.net blog 2015).

In Berkeley, after her marriage Sylvia Cranmer McLaughlin was busy as the mother of their two children, a faculty wife, and a generous hostess. She enjoyed her rose garden and the beautiful views of San Francisco from the terrace of their home. Then she learned that a third of the bay had been filled or diked off, and only 10 percent of the original wetlands remained. Less than six miles of shoreline was accessible to the public, and developers were planning to fill in 60 percent of what remained of the bay. Together with two friends, Catherine "Kay" Kerr and Esther Gulick, McLaughlin formed Save San Francisco Bay Association in 1961. It was one of the first grassroots environmental organizations and is credited with setting off a movement that spread across the country.

In the same year, 1961, Donald McLaughlin became chairman of the Homestake Mining Company board. Although there is no official record of it, undoubtedly Sylvia McLaughlin's environmental activities and her increasing national prominence were noticed and supported by her husband and affected some of his decisions for Homestake. Her work also influenced young environmental activists in the Bay Area, one of whom, Raymond Krauss, will appear later in the history of the McLaughlin Mine.

The plan for post-mining use as an environmental research facility named for Donald and Sylvia McLaughlin would be a crucial factor in obtaining an unprecedented number of permits needed for the mine to operate.

In 1977, its centennial year, Homestake Mining Company, with changed philosophy and management, would face daunting challenges as it met one of California's newest enterprises, the One Shot Mining Company.

CHAPTER 5

ONE SHOT MINING COMPANY

It's a school-boy fantasy and there still is some validity to it, because it happens.

—James William Wilder

"We called it One Shot and then we started. If we don't make it on this one, we're out of the mining business." In 1965 James William Wilder formed the One Shot Mining Company and leased the old Manhattan mercury mine in the Knoxville district of Napa County. His school-boy fantasy became California's highest gold producer of the 1900s, bringing twenty years of prosperity to Lake County, which until that time had been the poorest county in California. The mine was named for Donald Hamilton McLaughlin, a fellow Californian with a strikingly different life story. Both men were born in San Francisco, McLaughlin in 1891 and Wilder in 1924. Their lives were to converge in 1978.

Donald McLaughlin's father died when he was seven years old. His widowed mother was then employed by Phoebe Apperson Hearst as personal assistant. Phoebe was the widow of California senator George Hearst, a mining investor and one of the wealthiest and most influential men in America. McLaughlin became Phoebe Hearst's protégé and in 1914, when he graduated from UC Berkeley, Mrs. Hearst celebrated the occasion with a party for six hundred guests at her estate, Hacienda el Pozo de Verona, in Pleasanton. That fall, he began graduate studies in geology at Harvard University.

About ten years later, on April 4, 1924, James William Wilder was born in San Francisco, and baptized at Mission Dolores. His grandfather had come to San Francisco in 1860, and built a successful family business owning and operating tugboats that pulled or pushed barges loaded with goods in San Francisco Bay. Wilder graduated from Balboa High School in 1942, as he recalled:

> I was not really a bad student.... I was pretty good as far as things that I wanted to do that were important, I thought. Didn't agree with the school the whole time... [but] I got a diploma anyhow. (Wilder 1996, 3)

In October 1942, just after he turned eighteen, Bill Wilder enlisted in what was then the Army Air Corps, and served for two years in the Mariana Islands in the South Pacific, on B-29 aircraft. He left the service when the war ended and worked for his father's tugboat business in San Francisco until he watched some men working on dump trucks, and thought, "This has got to be the best business in the world." (Wilder 1996, 6)

He formed a partnership with a friend who bought a secondhand dump truck that Wilder operated. Business was good, and in a year he bought out his partner. He formed another partnership with three friends and bought more trucks. This pattern continued as he learned new skills, formed partnerships, made lease agreements, and bought equipment, then buying out the others and coming out ahead. Almost always, they left on friendly terms, winners all around.

IT'S THE CHICKEN TODAY AND FEATHERS TOMORROW

In the next few years, Wilder did a wide variety of work: hauling, construction, equipment rental, even excavation for a new casino in Reno. There were hauling jobs as far north as Weed, California, and a seven-month stint in 1952 building a sea wall and breakwater. An early job was hauling chromite ore from a mine near Livermore to a mill in San Jose. Then an offer came from Palo Alto Mining Company to do both the mining and hauling. He had two partners, Jim Hendren and Frank Turner. By then, he also had a wife and family and a home in Los Altos.

> I bought a shovel, a power shovel and an old loader, dozer, and I had the trucking equipment. So we took the whole job. It was based on production.... We had to lease some of the mines.... And the mill paid us. It depended on what kind of ore we produced and how much of it. That was the first education that boy, the chicken today and feathers tomorrow. (Wilder 1996, 10)

He learned of a mercury discovery in Redwood City, and he obtained a new contract.

> I saw a thing in the paper where a piece of property up in Redwood City...Farm Hills Ranch. They discovered this mercury deposit up there. That was Dick Dinely and Hal Pigott, who was an old mining fellow. So I went up there. I just had to see it.... I walked in and by gosh they had a loader working there doing some exploration. I stood around and finally.... They said, "Do you know anything about this?"
>
> I said, "Yes, that's what I've been doing, is mining, for a couple of years now."
>
> He said, "Well, have you got any equipment?"
>
> I said, "I've got a shovel and a loader and trucks."...
>
> ...[That was a] beautiful mine. Easy mining and some fantastically good ore.
>
> ...[They were using mercury] for mildew proofing for paints, for atomic plants. It was a heat exchanger on atomic subs. Gee, there's just so many uses. Mercury fulminate they were still using that for a blasting cap.... [The] chloralkalide plants were making the detergents and...chlorine compounds. It was used in them. The demand was good and the price was up. So we kept busy on that.... [That] must have been '55....
>
> ...That belonged to Andy Oddstadt. They call it Challenge Mining Company...because he said, "Boy, this is a challenge." (Wilder 1996, 16–20)

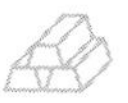

> The Challenge mercury deposit, also known as the Farm Hill No. 2 mine, Redwood City mercury mine and the Emerald Lake mercury deposit, represents one of the smaller mercury producers situated along the Coast Ranges of California....
>
> Although cinnabar had been known from the site prior to 1865, the "rediscovery" of the deposit during early 1955 created a minor "mercury boom" in the area with production of cinnabar and native mercury from the highly weathered serpentinite and silica-carbonate rocks. Numerous newspaper articles abounded with the "new discovery" and the potential of "grand riches to be made," even estimated in the millions by some people, which would be gained

> from the mining operations. Some even suggested that the deposit might rival those of the New Idria and New Almaden mines! (Dunning 2008, 1)

Now Bill Wilder had a new motivation.

> [There] was a little period of time in there around '56 that the guys from...Challenge, and myself, we formed a little partnership and we went up to Monitor Pass, the Lord Chalmers mine, the Kurtz, and Morning Star. (Wilder 1996, 28)

GO WHERE THE PRICE IS GOOD

Wilder began thinking that he wanted to own a mine himself.

> You read about mining and...it's sort of a little fantasy...you're going to pick up a nugget and there it is.... It's a schoolboy fantasy and there still is some validity to it in a way, too, because it happens.... I really started looking into this, you know, trying to learn everything I can about the mineralogy and how this occurs and what it occurs with.... I'd rather do something on my own....
>
> I was out hustling, trying to find some mine that looked like it was feasible to put the plant in to get production. So my first guess was up in Panoche Valley and down by New Idria.... We went through...every little old mine that had ever been recorded.... [Leased] the Hillsdale and the Chaboya...in San Jose...from old man Azevedo.... But it was no good. It was nip and tuck.... The amount of money you put in, you got just about that amount back. You never could get ahead. (Wilder 1996, 28)

He educated himself at home and during evenings in the library at Stanford University, studying records of the State Geological Survey.

> I'd go and buy some of the state bulletins and I'd research them. I got pretty good at researching.... It must have been '63–'64.... Mercury all of a sudden was up there at four or five hundred dollars. Then I said, "This is the place to go." Go where the price is good. (Wilder 1996, 30)

He found records of likely mercury prospects—the Gypsy, the Bluewing, the Coolwater, in the Stayton District, near Hollister. The claims had belonged to Raynor Knox, son of Richard Knox of Knoxville, and were now held by his daughter, Kathy Cushingham. Wilder went to visit her and learned that she had just leased to someone else. However, she thought he might be interested in another claim, the Manhattan, in Lake County.

> So I...started researching. And by God, yes, it's there. But nobody paid any attention to it because it had been inactive since 1906. It was still thirteenth in the country in production and it hadn't been run since...1905.... It was thirteenth in production and we're talking about things going through World War I and World War II and up to, say, '65. I went, "Holy smokes." (Wilder 1996, 32)

After further investigation, he decided it was indeed a good prospect.

> Came back up and it was one of those things when you think you know all the answers, that's when you really begin to realize how little you know. I came back and looked at it. After I came back about the fourth time...I began to get a pattern of how this thing was occurring.... I panned the dumps. Jeez, they panned beautiful.... I still have all the reports.... We did our own assaying on it and... [we] came up with enough values from the dumps to justify going ahead with the operation, building a mill and concentrating.... Now, if the dumps were that good, the mine should be better...[so] we went ahead and negotiated a lease. (Wilder 1996, 33)

Wilder and a former partner, Joe Matsumoto of Matsumoto Iron Works in Sunnyvale, formed a new partnership to work the Manhattan Mine. From his days hauling chromite ore from Mt. Hamilton, Wilder remembered seeing two old furnaces there.

> I went up there and I talked to an old guy, Roy Williams. He was about eighty-five then and we'd been friends for a long time. They had a little thing on this Arroyo del Mocho road out of Livermore that hits Patterson Pass if you turn to the left or if you go to the right, you go over to Mt. Hamilton, to the observatory. But out by that junction, there was a little store they called the Jot 'em Down Store. If you wanted something, you went in and told Roy, "Hey, when they're going down for supplies, will you have them pick up...this and this...." I think his daughter ran down every two days or something and picked up stuff.... I asked him about the furnace. He said, "God darn, there was [a] guy that bought those furnaces—was going to cut them up. Just last week he was talking to me."

> I hot-footed it over there and the guy…said, "Gosh, if I could get $700 for these furnaces, you've got them." I gave him…700, 750, less than a thousand, anyhow. And I bought those two furnaces and I said, "Now I own something."
>
> …They were sixty-five feet long. And they weighed about thirty-five tons apiece, thirty-seven tons apiece…up in a place where there's an eight-ton bridge.… I took the kids up there, took Billy and a bunch of friends of his. (Wilder 1996, 36–37)

It took all of Wilder's ingenuity to load each furnace onto a 1950 Sterling truck and maneuver it, sometimes inches at a time, down the mountain and on to the Manhattan Mine. Each trip took three long days.

> I still have the truck.… It's a 1950 Sterling that I had for transport. A Sterling is a classic truck. That's why I keep it.… It's better than the one that they got in the museum by a bunch. It's a better looking one and…it's more authentic.
>
> We had that Sterling and we got up there.… You only take one at a time. And it's a big load. It's a permit load; I had a permit for 102,000–108,000 pounds which is big-time.… I think we were greater than that. No way of weighing it out there. So we get the biggest permit we can get and we know we're going to be close anyhow.
>
> …So then we got down that…Patterson Pass Road…and there's a bridge that said seven, eight tons.… Boy, it was big time.… [We] were about fifty, sixty tons. So I said, "Whoa." Got under it and looked at it. It looks good and sturdy and it's got plenty of timber under it. I don't know where they came up with this rating but I said, "No, that's good for a lot more than that."
>
> Gerbers lived out there. And Gerber said, "I think that'll stand a lot more." We looked at it. The most we could do was break it. It's a wooden bridge so we'd have to rebuild it. So that's that.
>
> We just came on down and put the front axle on it real easy. It didn't even do nothing. Drove up and… heck, it made it all the way across.
>
> …Took it easy so it didn't bounce or anything.… Made it across; didn't even break a board in the bridge. I went all the way down and then it took me two days to get it into San Jose. Then I went back and got the second one. By then you got confidence, you know. It worked good. And we got both those furnaces in there.

> In the meantime we had Gordon Gould design this new plant utilizing those furnaces and all the stuff; we had a lot of stuff. Boy, we had a lot of equipment. What he did was to supply the engineering and we supplied all the equipment, motors and generators. I bought generators and rebuilt them myself because we didn't have any power up here so we had a good power plant. But we built big generators. That's something I knew; I'm good with my hands. I'm still working on engines. I never get away from it, you know. (Wilder 1996, 38–40)

When it was all finished, Wilder believed it was "the best plant in the country." On a personal questionnaire, he lists as his area of expertise: plant installation and design.

YOU FIND OUT WHO YOUR FRIENDS ARE AND WHO ARE THE SHARKS

He had a realistic attitude and modest goals for the enterprise:

> We called it One Shot and then we started. That was Tom Matsumoto's thing because he said, "Boy, this better be a one shot because if we don't make it on this one, we're out of the mining business. And that's that." That is a good name because everybody that goes into mining comes up with names like "World Wide Amalgamated Minerals Company" and they're a hole in the wall. Let's go to the smallest name we can think of where everybody is really doubtful. It's better to be little and get bigger than it is to be big and fail. (Wilder 1996, 50)

It took nearly a year to get the operation under way, installing a crusher, bins, and conveyor belts. When they first formed the partnership, mercury was selling at $740 a flask. Then the price began to slide down, to $400, to $300, finally down as low as $99 a flask. At that point, the Matsumoto brothers wanted out. Wilder bought out their interest in 1973, and formed a new alliance with Ed Wells, his bookkeeper, and Howard Wacasser, his right-hand man for maintenance. Tom Austin, an experienced miner, also worked for him. He bought a high-end trailer house and built on an extension so that he and his wife could live comfortably there. He bought other trailer houses for his sons and other employees, and gradually acquired a wide variety of other bargains: old San Francisco City buses, generators, trucks, and paint.

Whenever we made a buy on paint, we'd go to a flea market and buy all the paint we could find; that's why you end up with pink things with black stripes. The two-by-fours are black and the plywood's green or pink or yellow or—I mean, goofy colors. It looked nice.

[We] worked like dogs.... We just ran the heck out of it...took the best ore and ran it through the plant, loaded the pickup and hauled loads of mercury down to Mountain View.... We sold it ourself. We didn't have anybody handle it for us. It's better off to do it yourself.

Then you find out who your friends are and who was the sharks. (Wilder 1996, 53)

His operation was flexible; he produced whatever he could sell.

When the price got down, we built a...little cinnabar mill. An outfit called Lee Yuen Fung Company from Hong Kong came over and they contacted me. I was kind of amazed. Here's people from Hong Kong up here at the mine. You've got people from all over the world. We have people from the Almaden, Spain, and from Trieste and everywhere...I had a guest book that you wouldn't believe signed by these people from everywhere, all over the world.

Lee Yuen Fung Company.... They were really nice people and they said, "We need cinnabar over there for making jewelry boxes and stuff like that, and packaging certain aids, herbs. We preserve them."...He looked at some samples that we have of cinnabar.... So then we did that. We'd package it in five-gallon buckets and haul them over to the Sacramento Airport and air freight them to Hong Kong.

...We built a little plant to clean all the dirt out of it so it was a nice beautiful red, and not real purple. He didn't want purple. He wanted red, more of a scarlet color. Mercury at that time was selling for about a hundred dollars, ninety-nine dollars a flask which would be roughly a dollar and a half a pound. We were selling this stuff for twenty-two fifty a kilo in Hong Kong and they paid all the freight.... It was not a big operation but it helped pay the bills. So we did that. (Wilder 1996, 54)

The beautiful rock opened still another market locally.

Then we said, well, decorative rock. All that beautiful rock.... Sunrise Rock out of Sacramento come in and they started buying it...we

> built.... A crushing plant and screening plant to size it.... We made about six, eight different colors.... [We'd] run greens; we'd have a green product. We'd put it in the bins and then stockpile it with dump trucks. Then we're going to change to a white product. We had that white chalcedony, so we crushed that and stockpiled that. Then we ran the basalt calcines and we made a cocoa brown which is a beautiful brown color, better than this red rock. It didn't have that gaudy look to it. It was real subdued but beautiful. Then we mixed the green and brown by accident. That's a beautiful color, a tweed; so we made that. Nobody else had half these colors. We had a pretty good little business going on. (Wilder 1996, 54–55)

For a time, when the price of mercury went back up, he had a lucrative contract with the Mallory company to reclaim batteries. He became friendly with Mel Stinson, California State Geologist, who suggested that he get a gold assay. Despite his feeling that it was "like looking for elephants in the Mojave Desert," Wilder sent an ore sample to Martin Quist's Metallurgical Laboratories in San Francisco. There was no phone at the mine, so it was some time before he stopped at the lab and learned the results: about two-thirds of an ounce of gold in each ton of ore.

> But gold was thirty-five dollars an ounce so now we got twenty dollars a ton in gold. We've got mercury that is running thirty, forty dollars a ton. What are we going to be fooling around with gold at half the price? (Wilder 1996, 58)

Wilder was right; the price of gold had been fixed at $35 an ounce since 1934, when Franklin Roosevelt was president, but that was about to change. In August 1971 President Richard M. Nixon issued an executive order that ended the U.S. redemption of gold at $35 an ounce. John Livermore had found formerly invisible gold in Nevada; enormous machines could dig deeper pits than ever before, and mining companies were out hunting for gold again. In 1978 a geologist working for Homestake Mining Company took the first step that led to negotiations to buy the Manhattan Mine, and to develop Bill Wilder's venture into California's highest gold producer of the 1900s, named for Donald Hamilton McLaughlin.

CHAPTER 6

GOLD SO FINE THAT IT WOULD FLOAT

You're always doing research.
That's one of the things in exploration,
we're always looking at old reports.

—John Sealy Livermore

More than a century after Horatio Gates Livermore and his partners developed the XLCR/Excelsior mercury mine in Knoxville, California, his great-grandson John Sealy Livermore was working in Nevada as an exploration geologist for Newmont Mining Company.

In the intervening years, the Livermore family had become influential. In 1854 Horatio Gates Livermore was elected state senator for El Dorado County. His son Horatio P. Livermore acquired the Natomas Water and Mining Company and constructed the original Folsom Dam on the American River. Norman B. Livermore Sr. was a founder of Pacific Gas and Electric Company. Norman Livermore Jr. was state secretary of resources for Governor Ronald Reagan. Putnam Livermore was a founding director of the National Trust for Public Land. Mount Caroline on Angel Island in San Francisco Bay was named for Caroline Sealy Livermore in tribute to her successful effort to prevent commercial development of the island. The family ranch, Montesol, acquired in 1880, spread from northern Napa County into Lake and Sonoma Counties, encompassing the slopes of Mount Saint Helena and the historic Oat Hill and Corona mercury mines.

John Sealy Livermore, the son of Norman and Caroline, was born in San Francisco in 1918 and grew up spending holidays at Montesol. He graduated from Stanford University in 1940 as a geology major and had learned about the Getchell outcrop in Nevada:

> The story was that it was right on a main trail that a lot of prospectors and other people had passed for years, and they would bang off chunks of rock and crush them up and pan them, which was the standard way of prospecting in those days, and they never would get any "colors" of gold. This went on for quite a few years. But somebody had the bright idea much later in the thirties to send a sample in for assay. Well, it turned out there was gold in this rock, and this big mass of silicified material was all gold ore. But the gold was so fine that it would float out of the pan and not be detected, which was the reason it wasn't discovered until the thirties. That stuck in my mind, because I always thought, oh, that was kind of interesting. This is the type of gold that wouldn't pan, and maybe there were other similar occurrences that the old-timers might have missed. I always remembered this and I followed up on this later on. (Livermore 2000, 13)

John Livermore served in the U.S. Navy Civil Engineering Corps from 1943 to 1946. After the war he held a variety of jobs in Colorado, Florida, and Nevada, gaining experience as a geologist, a contract miner, and a prospector. In 1952 he was hired as a geologist for Newmont Mining Corporation and by 1960 he had inspected deposits in western states and Alaska, as well as Canada, Central America, South America, North Africa, and Turkey.

In 1961 he read a U.S. Bureau of Mines report from 1939 by O. W. Vanderburg that noted disseminated deposits in sedimentary rocks. He heard a talk given in Ely, Nevada, by USGS geologist Ralph Roberts, who had written a paper in 1960 that presented a new theory of thrust structures in northern Nevada. Livermore discussed with Roberts the theories about gold particles too small to be detected with the naked eye. This led him to begin painstaking fieldwork with his junior associate Alan Coope in northern Nevada in two areas known as the Carlin Window and the Lynn Window.

> So we decided just to systematically map and sample.... We covered a lot of ground very, very carefully, mapping and sampling and studying the rocks, and the alteration.... We were doing it very, very slowly...we might cover a mile a day.... We would map the rocks, and then take a lot of samples, and we would send them down to this [Harry Treweek's] assay office down in the Crescent Valley.... He [Harry Treweek] was a very good assayer. We had confidence in him because these samples we were taking were often very low-grade

samples, and we wanted to be sure that what he was reporting were true values, and not just spurious values. (Livermore 2000, 69–70)

Livermore's careful work in 1961 touched off a Nevada gold boom that recalled the California gold rush of 1849 and resulted in enormous profits for many companies. John Livermore remained a modest man, admired for his personal integrity. In tune with the other changes of the times, he supported in major ways causes other than mining: the preservation of historic Tonopah, Nevada, and the protection of wildlife and water resources in the Great Basin of southern Nevada. He was zealous in remediating the wastewater from the old mercury mines at Montesol.

John Livermore's proof of the concept of gold so fine that it would float guided Homestake geologists in their search for new gold mines. It would lead them to success at the site of Horatio Gates Livermore's XLCR mercury mine.

CHAPTER 7

SEARCHING FOR GOLD

Gold is considered to be the very best, the most exalted, the most important, the most beautiful, and the most noble of the metals.

—Lazarus Ercker, *Treatise on Ores and Assaying*, 1580. (Translated from the German by Anneliese G. Sisco and Cyril Stanley Smith)

In 1975, in the executive offices of Homestake Mining Company in San Francisco, consultants from Boston, the Athens of America, presented their conclusions. After extensive visits and interviews with employees at the company's far-flung operations, they arrived at the same decision as George Hearst, an unschooled miner from Missouri, had a century earlier: the best thing is to locate a good gold mine. Then the search began.

Panning for nuggets in Deadwood's creeks was far in the past; John Livermore's discovery of gold too fine to be seen, and the types of formation that held the invisible gold, had changed everything. One aspect of the Homestake reorganization was to hire a chief of exploration for the United States. James Anderson, a Harvard-trained geologist, was hired in November 1975 as executive vice president for exploration, with a budget of $60 million over five years. He welcomed the intellectual challenge of finding a new deposit.

Anderson was born in Aurelia, Iowa, in 1935, in a farming family. When his father joined the Navy in 1941, Jim and his mother, a schoolteacher, moved to Eureka, Utah, where they stayed after his father's discharge in 1945. Jim studied geological engineering at the University of Utah, and then at Harvard, where he was advised by one of Donald McLaughlin's former students, Hugh McKinstry. He worked summers with surveying crews and in the mines around Eureka. From 1960 to 1967 Anderson was an exploration geologist at Kennecott Copper Corporation's research facility in Salt

Lake City, developing exploration procedures for copper deposits. In 1967 he became vice president of exploration for Occidental Minerals. He recalls,

> [I] was involved in all the business aspects of exploration plus development, up to the point where you turned it over to operations, the people that will build and operate the mine.... It was exciting. But then a more exciting opportunity came along, to join Homestake and work with Don McLaughlin and Paul Henshaw.... I enjoyed going from a bigger corporate environment to a smaller group where everybody worked together closely as a team, you knew everybody, and you knew where you could go to get help directly without having to go through layers of staffs and people to find somebody. (Anderson interview, KMD vol. 1, 15, 18)

James Anderson initiated the search that led three years later, in 1978, to the Manhattan mercury mine in Napa County, California.

> From my experience...in thermal dynamics and physical chemistry...I had deduced that gold deposits ought to occur in hot springs...[but all] the literature that I had read had indicated that gold did not concentrate in hot springs, and that was very discouraging.... [Late] '77, October or so...I took a vacation for a couple of weeks in...the San Diego area, and...I spent every day at the UC [University of California] San Diego library...researching all the obscure publications I could to find out whether or not gold occurred in hot springs.... [At] Carlin [Nevada] the gold was micron gold and...you couldn't see it without a high-powered microscope...if nobody had ever assayed for gold and it was too fine-grained to pan, then it didn't prove that gold was absent in hot springs deposits.... I came back and discussed the concept with Don McLaughlin...and Paul Henshaw, and suggested that we consider implementing a major reconnaissance....
>
> Homestake had a fabulous set of files at Lead...representing a hundred years of investigations of gold deposits...around the world.... So what we should do is go up to the Homestake mine...[find] out if any of the gold prospects might be associated with hot springs. (Anderson interview, KMD vol. 1, 23–25)

An exploration office had already been set up in Reno in 1975, staffed by exploration geologists Ken Jones and Donald Gustafson. Recently hired, Gustafson was born in Illinois in 1938, and grew up in the house that had been the family home for three generations. His only connection with earth sciences then was working in the gravel pit of his father's ready-mix business. He earned both bachelor's and master's degrees at the University of Colorado, then worked for the USGS and for ten years as a mine geologist for Anaconda Company in Nevada and Montana before joining Homestake in 1975. He says of exploration geology that it "really all boils down to a lot of common sense and a lot of patience." (Gustafson interview, KMD vol. 4, 16)

GEOLOGISTS ARE ALWAYS OPTIMISTIC

Working out of the Reno office, he explored in Nevada and then at Bodie, California.

> Bodie was one of the prospects that we did quite a bit of work on, that really triggered the thought of the mercury-hot-springs-gold association.... The old town now is a California historical site. It is well preserved and...in such good shape. At Bodie...there are chalcedonic veins that do contain some cinnabar along fractures and if you reconstruct the geology...the chalcedonic veins with the cinnabar were at a higher elevation than the gold mineralization.... The gold...would have been deeper than the...mercury...when the deposit formed.
>
> So that—plus some other features from other deposits that we looked at—kind of made a person think maybe you could go look at the mercury mines where you have chalcedonic quartz and cinnabar at the surface, and drill through that zone into a gold-bearing zone below the mercury mineralization.... Bodie has a soft spot in my heart because it did help develop the mercury-hot-springs concept.... It was a new concept of really where to look for gold deposits. (Gustafson interview, KMD vol. 4, 24–25)

Jim Anderson assigned geologists Gail Hansen and Joe Wargo to go to Lead, South Dakota, to examine the files there, and to send the promising files to Reno. In the summer of 1977, Don Gustafson looked at a report from half a century earlier.

> The Cherry Hill-Wilbur Springs gold-mercury deposit which Homestake had looked at in 1925, 1926. A fellow by the name of Bill Yates... looked at this property. It had been mined as a gold mine. It had been mined as a mercury mine. And there were active hot springs in the area...Colusa County, California. And their prediction, from that 1926 report based on sampling that had been done, was that there

was open-pittable 10 million tons of .08 [eight hundredths of an ounce of gold per ton] on that property. In 1977, that gets one's interest up. Here's a Homestake geologist; he must have been relatively competent and if he thinks there's 10 million tons of .08 there, that you can mine with an open pit, then a person better go look at it. Plus, it had the makeup of gold mineralization, mercury mineralization and hot springs activity all in one place. It was in sedimentary rocks which would be favorable for widespread mineralization. I looked at the report, middle of 1977, and with the experience at Bodie and the thought about this mercury-hot-springs association—(Gustafson interview, KMD vol. 4, 26–27)

Gustafson went to Colusa County to find property owners there.

Mr. and Mrs. Weightman...had 150 acres, which was essentially the mineralized portion of the area. As you go into the Wilbur Springs hot springs, near the spa area, there's a little bridge and they had a phone there; I called them from that little bridge to get their permission to go look at the property. They were pretty hesitant because they'd had some mercury miners in there and they kind of botched things up.... But I finally...convinced them...looked at the property...it was exactly as Bill Yates had said. It was sedimentary rocks, mercury mineralization, some alteration, hot springs activity, just everything that he had said was there. And you could actually go back to the original sample sites which were marked with little wooden blocks; the numbers were still there.... Well, I took at least forty samples. Went back to Reno, had those analyzed in the lab, averaged up the forty samples, and they averaged .08. And I thought, "Hm. The guy must have known what he was doing."

So, now we're into October of '77 and every fall, the Homestake exploration group had their budget presentations. So, I was really excited and was going to present my mercury-hot-springs concept. I had an idea, had a concept. But now, and more importantly, I had the Cherry Hill property that proves that you can have mercury-hot-springs and gold in economic quantities all in one place. So, it just couldn't go wrong.

So, had the meeting in Lakewood [Colorado]...October of '77. And all the exploration group was there: Gail Hansen, Jim Anderson, Joe Wargo...Ted Rizzi...maybe twenty-five people. Anyway, I presented my concept of the mercury-hot-springs gold association

> with a model cross-section showing how it could happen and then presented the data from my initial examination of Cherry Hill.... So, anyway, to make a long story short, we tied up Cherry Hill.... And then we had to go through the Colusa County commissioners... to get permission...it was the end of '78 before we started to drill. (Gustafson interview, KMD vol. 4, 28–32)

The Cherry Hill project, though, was a disappointment, as Homestake geologist Dean Enderlin recalls:

> We did try hard to make a mine out of Cherry Hill, not just in the late seventies, but on up into the late eighties, and never could. There were probably three successive periods after the 1978 drilling program that we went back to Cherry Hill to try to understand it. Every time we went, exploring different concepts, different ideas, and finding a bit more resource, but never enough to make it into a mine. It's still enigmatic that that deposit has seemingly no root. Very high grade at the surface. You can chip away at the gold veins there and see gold with the naked eye...Cherry Hill was a very tempting deposit, and very frustrating. (D. Enderlin interview, KMD vol. 8, 33)

Enderlin tells that the Homestake geologists learned a valuable lesson at Cherry Hill: the limitations of the rotary drill.

> The original exploration drilling that was done at Cherry Hill around '77 and '78 was done rotary. Rotary drills as opposed to core drills are very fast, very cheap; they generate a big sample, and actually in terms of drilling, it's a real quick way to identify a resource very rapidly.
>
> One caveat in using a rotary drill is that under wet conditions in rock formations such as we have, you can concentrate the gold.... Homestake discovered that fairly late in the game at Cherry Hill. At some point during the course of exploration, after announcements had been made, after the momentum had gathered to move the Cherry Hill project towards being a viable mine, someone decided to come in and diamond drill, using core drills, to compare some of the rotary results. And the results showed that wherever a rotary hole had been bored, and a core hole, or diamond drill hole, had been bored next to it, the results didn't match well, that the core was typically consistently of lower grade. And that would be consistent with

> that flaw in rotary drilling techniques, what we called "panning," literally.... Your rotary drill is acting much like a gold pan, washing away the barren stuff, the fine stuff, and concentrating the goodies.
>
> ...[We] can look back and say well, yes, we learned our lesson early on, on a much smaller deposit, and fortunately we were able to make a good thing out of it by discovering the McLaughlin deposit in the end.... [Geologists] are ever the optimists.... We wouldn't be employed very long if we weren't. The last thing you want on your staff is a pessimistic geologist. (D. Enderlin interview, KMD vol. 8, 31–33)

Meanwhile, Don Gustafson was busy looking at more prospects, still hoping to prove his theory of gold associated with mercury and hot springs.

> I would go out and do the initial examination, take a few samples by myself.... I would spend maybe two days there. And then if I saw one that looked interesting to me, Joe [Strapko] and Tom [Timken] would come and they would spend two or three weeks. They would do some detailed mapping and sampling of the property. That's how that worked. And in February of '78 is when I first went to the Manhattan. (Gustafson interview, KMD vol. 4, 33)

Don Gustafson's optimism would be tested again at the Manhattan Mine site.

CHAPTER 8

THE FIRST STEP

Manhattan was just a joy to look at.

— Donald Gustafson

Homestake geologist Donald Gustafson walked onto the Manhattan Mine property on February 16, 1978, taking the first step on a journey that would lead to the development of California's most productive gold mine of the twentieth century. It would involve, among many others, a mining engineer born in Chile and another born in Cuba, a metallurgist from Australia, a construction engineer from New Zealand, a geologist whose grandfather had come to Lake County a century earlier, a mechanical engineer from the iron mining country of northern Minnesota, ceramic workers from Germany, gold processing experts from South Africa, and a planner whose ideas were shaped by life in San Francisco in the 1960s when "ecology" and "the environment" first became common terms. This effort would culminate five years later with a stack eight feet high of 327 permits from various counties, water districts, and air quality districts.

There are differing accounts of the first meeting with Bill Wilder. Donald Gustafson recalls it this way:

> And I drove in there on February 16, 1978, to look...at the Manhattan mercury mine which was owned by Bill Wilder who was living on the property in a trailer.... He, his wife Kay, and his men; he had maybe two or three people that were working for him.... I met Bill, a very lovable, happy-go-lucky individual who I really admire and like.... I told him I was interested in evaluating the mercury potential of his property.
>
> Of course when you say Homestake Mining Company to anybody, they immediately think of gold and when somebody says, "I work for

Homestake Mining Company and I'd just like to evaluate the mercury potential of your property," the guy looks at you like, "Hm. This guy's really not telling me the straight scoop," which I'm sure Bill thought. But anyway, "Yes, no problem; I'll show you around."

And Bill loves to talk. I spent two days there and I probably talked to Bill fifteen hours a day.... [He'd] take me around and he'd show me this and say, "Well, let's go over here and look at this." Well, he was taking me to more or less the outer fringe areas of the property which is where he was mining the mercury.... I wanted to look more in the central part where the hot springs activity was and the quartz veining.... But I kind of went along with Bill and let him give me the tour. And we hit it off pretty well together. I spent a couple of days there and took a bunch of samples, made a quick geologic sketch map of the area...went home and got the samples analyzed and lo and behold they ran a high of a third of an ounce of gold.... So that's what got us started on the Manhattan mercury mine evaluation. (Gustafson interview, KMD vol. 4, 33–34)

Wilder's recollection is different. He recalls he was busy producing decorative rock:

Don was the first guy who came in and said, "Could I look around?"

"What do you want to look for?" We were busy working on this rock plant and it was a miserable day.

He says, "Oh, just minerals, you know."

"I'm busy right now." We were...rebuilding a crusher that I hauled up...and so we said, "Have at it."

He came back and said, "Gee, I'd like to come back again."

"Good, come back."

So he came back and then next thing I know, he came down with Ken Jones. They said, "Gee, we'd like to buy the place." (Wilder 1996, 55, 57)

In another recollection, Wilder says he told Gustafson, "Just stay out of our area. We're busy. We don't have any intentions of leasing to you. We don't have any intentions of selling to you. So you understand that you don't do a lot of work for nothing" (Wilder 1996, 57).

Although Wilder doesn't mention it, a team of exploration geologists did a lot of painstaking work at the Manhattan site between that first visit in February, and the later one by Ken Jones. Gustafson recalls,

> I took Tom Timken and Joe Strapko and said, "Hey, you guys go spend two or three weeks at Manhattan and map and sample that area," which they did. At the end of like a three-week period, they had over a hundred samples.... We had those analyzed and there was a central core which later, in the mining days, became what they call the central zone, that averaged greater than .1 ounces of gold per ton on the surface. And there were a lot of scattered occurrences out of that zone. So it was developing into a pretty good-sized surface gold expression.
>
> Manhattan was just a joy to look at. It has to be one of the best exposed, zoned gold deposits that ever was. It was just a type example...of the mercury-hot-springs-gold association; a classic example. Just all laid out before you....
>
> ...[It's] a matter of timing and then economics.... And with Manhattan, they all came together just right. (Gustafson interview, KMD vol. 4, 36–37)

Gustafson says that, on one of their visits, Wilder told him that he had a copy of the classic Becker report, which was often quoted by geologists working in the Knoxville District. In the nineteenth century, geologist George Ferdinand Becker was sent out by John Wesley Powell, director of the USGS, on two expeditions to northern California. The earlier examination, in 1885, was reported in Bulletin Number 19, entitled "Notes on the Stratigraphy of California."

Results of the later expedition were told in Monograph 13, titled "Geology of the Quicksilver Deposits of the Pacific Slope, with an Atlas," by George Ferdinand Becker, published by the Washington Government Printing Office in 1888. This report is undoubtedly the one that Wilder had:

> The Manhattan and Lake mines, which are contiguous claims, lie to the south of the basalt area. It is somewhat remarkable that scarcely a trace of serpentine exists near these mines. The surface soil here and also near the Redington mine contains cinnabar, resulting from the erosion of croppings, and accompanying the cinnabar is free gold, which may be found by panning the soil. There is no doubt that

> a part of this gold, if not the whole of it, was originally contained in pyrite. (Becker 1888, 282)

YOU'RE ALWAYS GOING TO BE OPTIMISTIC IN EXPLORATION

Joseph Strapko had been with Homestake for less than three years when he was assigned the job of sampling and mapping the ground at Bill Wilder's One Shot Mining Company. Strapko was born in Watertown, Wisconsin, in 1951, and majored in geology at Beloit College. As an undergraduate, he worked on two National Science Foundation projects, and another with a mining company in Washington, plotting data from a survey of stream sediments. After graduating in 1975, he wanted to work again in mining, so he went to Denver to look for a job. He recalled that his first tries were unsuccessful:

> It's where I got an idea that geologists were very nice people. They really gave me a lot of help, and they would send me to somebody else, and that person would say, "No, I don't have a job." And then that person would try to think of somebody. And finally, "Wait a minute. I think those people over at Homestake are looking for somebody." (Strapko interview, KMD vol. 6, 257)

Strapko was hired as a field assistant in May 1975, and spent the summer in Breckenridge, Colorado, on a drilling project. From there, he went to the Homestake Mine in Lead, South Dakota, for a stint as a stope geologist, and then to Calumet, Michigan, where he logged 50,000 feet of core. He recalls,

> Logging core is the logical extent of exploration, where you come up with ideas and you say that something—an ore deposit—should be here. Well, you drill your hole to find out if it really is, so that part of it is exciting. The part of going in and logging core eight hours a day, that gets to be rather boring... [even though] it was a good experience. (Strapko interview, KMD vol. 6, 263–64)

Note: "core" is the cylinder of rock that the drill produces; the "log" is the record of its location.

In late 1977, Strapko and Ray Wilcox flew from Michigan to Homestake's Reno office and met with Don Gustafson to discuss the exploration activity in California. They welcomed the transfer; in upper Michigan winter was coming on. After spending a few days in the Reno office to make plans, they went to Colusa, California. Homestake gave them a Chevy Blazer, and they looked at Cherry Hill, the Abbott Mine, and a couple of the other mercury mines. Strapko recalled,

> I was doing...not detailed mapping, but a little bit more than initial reconnaissance.... After they found the gold deposit, there was an attempt to make it sound as though this was a very high-tech effort,...but it's not that sophisticated. That's the type of talk and something you send out for stockbrokers. You project this high-tech image so they believe that you'll continue to find things. If they realized how hit-and-miss it was, you probably could never raise money for a mining venture.
>
> [Y]ou're always going to be optimistic in exploration; with pessimism you'll never find anything because no project will ever look good enough. (Strapko interview, KMD vol. 6, 265–66, 269)

Strapko recalled how they approached landowners:

> We didn't get permission [from the landowners] in most cases, unless there was somebody there or unless it said, "No Trespassing. Contact so-and-so."...If you talked to a landowner you, of course, had to tell you were a geologist and you were doing mineral exploration, but you wouldn't say, "I'm looking for gold."...You would say, "I'm looking at the mercury prospects here," and leave it at that.
>
> But, you know, if there was nobody there, you would just go in...and look at the rocks and chip off some and take a [five-pound] sample.... We would actually drive them back to Reno.... There was a lab right there where we could drop them off at.
>
> ...[T]hen, if you find gold values,...you can go back and do a more detailed job and sample everything: take good channel samples, find out exactly how much gold is in exactly what type of rock.... [T]hat takes a lot more time. (Strapko interview, KMD vol. 6, 270–72)

Strapko recalled his first visit with Don Gustafson to the historic Manhattan mercury mine, now owned by the One Shot Mining Company, in February 1978:

> [T]he owner there, Bill Wilder...was on the property the whole time, so that was one of the areas where you couldn't really trespass.... Don introduced me to him...[he] said it was fine, that we could... take a look around.

> I picked up three samples. One was silicified volcanic; one was silicified sediment; and one was volcanic that had a big quartz vein running through it, with pyrite in it.... [When] we got the sample results back from...the silicified ones...those two were .07 ounces per ton.... And the one with the vein in it was almost a third of an ounce.
>
> ...Those were high values.... [S]o then we did the next step... [which] took about two or three weeks. Probably took a little bit longer than it should, because we kept ending up talking to Bill Wilder a lot.... Bill is a real talkative guy.
>
> ...We just went out, did some basic mapping, and sampled, made sure that everything...that looked different was sampled.... And the area was fairly well covered; nothing was going to slip through the cracks there.... It ended up with an area that was about a mile long and 800 feet wide.... That's a big target.... [W]e took over 100 samples.... [T]he average value in the samples was a little over a tenth of an ounce per ton, which is very good grade for an open pit. (Strapko interview, KMD vol. 6, 274, 275)

Between February and April 1978, Joe Strapko and Tom Timken were busy at the Manhattan Mine, taking samples and mapping a potential ore body. Homestake set up a trailer on the property where they updated the maps at the end of the day, and where they often spent the night. Wilder also let them use a trailer where they set up a drafting table. They picked up and chipped rocks to get more than a hundred samples over an area that was a mile long and eight hundred feet wide. Strapko says,

> if you're just detail mapping and rock chip sampling, you don't need permits. But we wanted to get some idea that the gold values weren't just at the surface...using an air track drill...what they use in quarries and open pits to put a hole down really quickly—a three, four-inch diameter hole.... We weren't going to go over fifty feet. (Strapko interview, KMD vol. 6, 284)

After the samples were analyzed, an ore body was delineated, the central core of which was rich enough to justify going ahead, and this required permits. Homestake attorney Dennis Goldstein was assigned the task of obtaining them. He was born in Providence, Rhode Island, graduated from Brown University, and was a student at Stanford law school in 1970 when the California Environmental Quality law was passed. His first job was with the California attorney general's office, and he worked on a case in which Napa County passed an ordinance to permit companies to develop geothermal projects. His experience and connections in the county and the state were fortuitous and he was hired by Homestake in October 1976.

When Don Gustafson invited him in 1978 to visit a prospect in Napa County, his first reaction was that it was a waste of time, that no one could open a mine there. He was thinking of the valley with its famed wineries, but Gustafson showed him that this was entirely different. Goldstein recalled his surprise:

> I think we drove up the Berryessa Road and...the last telephone we came to was at the north end of Lake Berryessa.... I was amazed that there could be any place in Napa and Lake Counties...where you could not be within twenty miles of a telephone.... It was so remote and had been so worked over and disturbed and messed up by generations of miners before that it immediately became apparent to me that it shouldn't be an environmental problem. (Goldstein interview, KMD vol. 3, 170)

First, they met with James Hickey, the planning director of Napa County. Before moving to Napa in 1970, Hickey was regional planning director for the Association of Bay Area Governments, composed of nine Bay Area counties and ninety-three cities, including Berkeley, Oakland, San Francisco, and San Jose. By 1978 he had learned to meet with citizens, city councils, planning commissions, and county supervisors. He knew that "whenever you get a political agency, you have organized groups who either want you to do something or want you to do nothing" (Hickey interview, KMD vol. 4, 108).

Gustafson and Goldstein took with them an aerial photo of the area, to show how disturbed it was already, and also a written proposal for what they planned to do. Three years earlier, California had passed California's Surface Mining and Reclamation Act of 1975 (Public Resources Code, Sections 2710–2796) requiring

1. The reclamation of mined lands will be carried out in such a way that the continued mining of minerals will be permitted;
2. The adverse effects of surface mining operations will be prevented or minimized, and that the lands will be reclaimed to a usable condition which is readily adaptable for alternative land use;
3. Residual hazards to the public health and safety will be mitigated to the maximum extent possible; and

4. The production and conservation of minerals will be encouraged while giving consideration to values relating to recreation, watershed, wildlife, range and forage, and aesthetic enjoyment.

> The Act's requirements apply to anyone, including government agencies, engaged in surface mining operations in California (including those on federally managed lands) which disturb more than one acre or remove more than 1,000 cubic yards of material. (SMARA Code Enforcement)

Napa County, however, had not passed an ordinance authorizing mining activities there. Goldstein recalled that the Napa County planners seemed bewildered because they had never been approached by a mining company to do exploration for gold. It was a remote part of Napa County, and most of the people in Napa didn't even know that was a part of Napa County. They worried that the mine would bring development costs to the county, and wanted to be sure that Homestake paid its own way (Goldstein interview, KMD vol. 3, 188).

Permission was granted for limited drilling, and to use existing roads, with minimum disturbance of the land. When Joseph Strapko returned to the One Shot Mine to do deeper drilling, he warned Bill Wilder about his decorative rock business:

> [A]t that point, the cat was out of the bag about the gold. He had been crushing up some of this rock, this chalcedony, and selling it as decorative rock, and that was where the best gold values were, so when I came back in December he asked about that, and I said, "Well, Bill, the stuff that looks the nicest, that's the stuff that's got all the gold in it. A lot of that stuff is running over an ounce per ton. That's quite a bit of gold. Don't sell any more of that." (Strapko interview, KMD vol. 6, 278)

A MECHANICAL GENIUS

Dennis Goldstein remembered going back to Wilder's place:

> Okay, well, Wilder is sort of a mechanical genius.... Wilder had this enormous "junkyard," I would call it, on this property. And he'd be very offended if you called it a "junkyard" in his presence, though. It wasn't a junkyard to him.... It was his "boneyard," as he'd call it. And I remember going up there for the first time.... Don [Gustaf-

son] took me up there. As we came around a bend on this dirt road there were acres of old machines...cars and all kinds of mechanical devices.... I think I counted eight...old San Francisco buses that still said "8 MacAllister" on them. They were the buses with the pancake engines under the floorboards of the bus...the buses had no nose or hood.... And so many other vehicles...each one of them had a use for Bill. (Goldstein interview, KMD vol. 3, 175–76)

One example of Wilder's ingenuity:

I remember one particular episode. We had an exploration drill on this property. And it broke down. And Don Gustafson...got on the phone and...called the Longyear drilling company in either Phoenix or Tucson...to obtain a replacement part for the drill.... It was going to take two weeks to get this particular part....

Before you know it, here's Bill Wilder. And he goes over to the drilling trailers and he studies the specification manuals that came with these drills, you know, like a Chilton manual for a car. And then he went down into his boneyard and he stood on the hill there for a while and he took some part out of one of these old vehicles...and he took it into his machine shop and he machined it; and in a couple of days, he made a custom part for this drill and they put it in and got that drill running again. That's the kind of guy Bill Wilder was. He is a very, very smart man and from my way of thinking, a mechanical genius: a guy who had the vision to see how things fit together spatially and how they should work together. (Goldstein interview, KMD vol. 3, 176)

Another example of Wilder's creative talent:

[H]is facility...was an amazing jerry-rigged facility. The first time I went up there, I could hardly believe my eyes. He was taking ore, or millfeed, whatever he was feeding into his kiln at the time. And he was mixing it into a slurry. He had a large trough outside of this facility with a couple of water hoses coming in. Somebody was shoveling material into it. Some material was also coming in on a conveyor belt. He was mixing all this together with a small old Evinrude outboard engine, the kind that I had seen on the back of small skiffs when I was a kid and went fishing with my dad. And this little propeller was turning and mixing the material that was fed to process.

And I went into this mill; he called it a rotary kiln.... I called it a retort and he was careful to correct me and say, no, that's not a retort. That's a rotary kiln. I don't know what the difference is to this day. But if Bill said there was a difference, I'm sure there was. This rotary kiln had an exterior lining on it for insulation. And it was made of flattened out, large-size Saffola oil tins...turning round and round: saffola, saffola, saffola, over and over. The slurry that was coming up and falling into the kiln had come through the trough with the little Evinrude engine on it and onto a conveyor belt that was stopping and starting automatically. And Bill pointed out to me that he had taken the electric eye of the type that opens the supermarket door. They had junked it. It wasn't working right so he took it and fixed it up and he installed it. And when the millfeed...conveyed into the entry to the kiln filled it up, material would...build up...and interrupt the light beam from this electric eye and turn off the conveyor. That homemade device is a testament to the kind of man Bill is. (Goldstein interview, KMD vol. 3, 176–77)

RIVALRY BUT REALLY GOOD FRIENDS

Bill Wilder had his own recollection of the early days when Homestake crews first began working at the One Shot:

The crew that Homestake had and the One Shot crew. It was like a big family. There was a bunch of rivalry but really good friends. I mean, they were real close.... It was almost like a bunch of guys that join the army. You get really close together and look out for each other. If a guy needs something, he's got it, right now. He's got a lot of support, like having a lot of brothers. It was good.

I was over in Adelaide in Australia and talking to John Roberts. And he said, "God, I've got a guy in here that was up there on exploration with Homestake."...And he went and got the guy, a geologist, and the guy said, "Gee, that was the best job I've ever been on any place in the whole damn world." He said, "Hey, that was the neatest thing because everybody got along so good. If a guy didn't work, he was gone in less than a week. Everybody worked like heck.... It was ran so good and it was so much fun. We had fun while we were working, worked really hard." (Wilder 1996, 67)

MY MOMENT OF GLORY

In April 1978 Joseph Strapko led the Homestake directors on a tour of the Manhattan project. He had been with the company for just shy of three years, and it was a stressful occasion for the young geologist, as he recalled:

> At that time, . . . they had outlined an area about 800 feet wide and a mile long. Well, . . . I did a calculation—taking it out only to 100 or maybe 200 feet, there was about 150 or 160 million tons. . . . [I]t was looking good. . . . [E]ven if there was large amounts of internal waste, there was a lot of good-looking rock there that had the potential of becoming a large gold mine. (Strapko interview, KMD vol. 6, 280)

Harry Conger, president of the company, began to grill Strapko on the samples he had taken:

> "This five feet wide. Is that a true thickness?"
>
> And I was pretty sure it was a true thickness. . . . And he kept asking me. . . . I got really flustered. . . . I said, "Well, I'll go get a tape measure." . . . And he grabbed me by the arm.
>
> He said, "Oh, come on. We're just giving you a hard time here."
>
> . . . I was young enough where I got flustered. . . . But anyway. . . . I think over all it was a pretty good tour. . . . I had my moment of glory there. (Strapko interview, KMD vol. 6, 280)

Homestake officials were convinced that it was a good project, and now the property had to be leased or bought from Wilder and other neighboring landowners.

There was still a long road ahead. Even after the land was acquired, Homestake had to overcome fears of local farmers, opposition by environmentalists, and a major surprise in the composition of the ore, before, as Don Gustafson predicted, "they all came together just right" (Gustafson interview, KMD vol. 4, 37).

CHAPTER 9

SECURING THE PROPERTY

I've got my whole life in this thing.

—William Wilder, owner, One Shot Mining Company

From April to December 1978, Homestake attorneys worked to secure the property for a gold mine from neighboring landowners. Dennis Goldstein outlined the usual procedure:

> Well...what we like to do at the beginning of a prospect or a project is to take a lease with an option to purchase so that if you find something of value pursuant to your exploration, you can exercise the option and buy the property without having to go back and renegotiate the purchase price.... The big interest for the lessor or the landowner in this case was to get the monthly rental payments; sometimes we'd call them rental and sometimes we'd call them advanced royalties.... We were entitled to go on and mine under the lease and we would pay a royalty that was more or less the market rate for gold-type operations at that time.... The landowner would prosper if the company prospered.... And of course, if we found something and ultimately produced it, as we did on this property, the landowner would receive a royalty which is a percentage of value derived from the sale of the product.... I think everybody would agree, that turned out to be very lucrative for all of the landowners. (Goldstein interview, KMD vol. 3, 183)

GO OVER THE PAPERS TO THE LAST EYELASH

In 1994 Goldstein recalled the negotiations for the McLaughlin Mine:

> Across the ridge where the water reservoir now is and where the north end of the mine is, there was another landowner...Robert

Kauffman. This [Kauffman] family had been on that property for some generations.... Bob's grandfather was a principal in a lumber business in San Francisco.... And in old pictures...of Sutter or Market Street...you sometimes see a big storefront that says either Kauffman and MacArthur or MacArthur and Kauffman. This was a successful business. And they were in the business, among other things, of supplying mine timbers to the mercury mines up in the area of what is now The Geysers and the McLaughlin mine.... And he [Bob's grandfather] began to acquire property there...they had five or seven thousand acres under their control.... They had some mining claims and some patented ground and some leases from the Bureau of Land Management. They were running cattle.... They were sort of ranchers. But Bob's individual history is really not that of a rancher. He was a businessman.... He wanted to subdivide and sell off the lands...that are now part of the McLaughlin project. He had this idea that he would build a lake, not too far away from where we ultimately put our water reservoir. (Goldstein interview, KMD vol. 3, 173)

Kauffman had a partner who was glad to sell his property:

And he found himself a partner whose name was Swanson from whom we acquired ultimately some property...about twelve or fifteen hundred acres.... And it included a piece up on the ridge where the county line was.... Swanson had a little air strip and some stock raising ponds up there.... [Swanson] was the easiest guy to deal with...he sold the property to us outright.... He had a land-leveling business down in San Luis Obispo County.... He didn't want a royalty and he didn't want to be involved in any mining.... [He] just wanted out.... so that's how we got that 1200 acres. (Goldstein interview, KMD vol. 3, 184)

Kauffman owned some key property for development of the mine:

Most of Kauffman's land was in Yolo County, but he owned land that crossed over the line into Napa County. And just to put that into perspective, the McLaughlin pit itself stretching from the south to the north is a series of patented and unpatented mining claims and the three most northerly of these were controlled by Kauffman. All of the others we purchased from Wilder.

> Bob also mined a little bit. He never mined to the same magnitude as Wilder. But on the properties that the Kauffman family then controlled—most of which had come to Bob from his grandfather—there were several old mercury mines. Bob and his boys would go out and work these mines on the weekends and sometimes for longer. But this was pretty much a function of the price of mercury. And there had been a period, I think either in the late '60s or the early '70s, when the price of mercury was quite high, as high as $500 a flask. At that time, Bob and his family found it profitable to work the property for mercury and... so they knew something about mining. Bob was a very mechanically minded person who was good with his hands. Bob could mine, Bob could build, Bob could grade.... He was a very different personality from Mr. Wilder. Wilder is a very trusting individual. You could build up a level of trust with him.... Kauffman, on the other hand, had built up... a distrust for... large organizations, governments, corporations. (Goldstein interview, KMD vol. 3, 179)

He was emotionally attached to his property, and was cautious because there was family lore about a long-ago sale that was a mistake:

> He was a perfectly wonderful individual other than for that one trait that got in the way of doing business with him. He had a wonderful sense of humor and an enormous attachment to that land that had been in his family for so long.
>
> And if he told me once, he must have told me ten times about something his grandfather did. His grandfather, although he owned all that land, sold the minerals underneath, what later became the Reed Mine. And I think that the family lore for years after that was that that had been a mistake and that much money had been lost by selling the Reed Mine.
>
> ...Bob had a couple of lawyers involved on his side. One was an old family friend... [who] stayed on as an advisor to Bob throughout this entire period.... His name was Paul Goulart and he practices law in Sacramento.... Then he had a wonderful lawyer who lived in Woodland. John Young was his name. He's not with us any longer... he was a country lawyer... not skilled in mining necessarily but a smart man.... Paul and John would go over the papers to the last eyelash and we killed a lot of trees doing that deal. There was paper and

more paper and more paper still. But we finally got it done. (Goldstein interview, KMD vol. 3, 181)

The final arrangement with Kauffman satisfied both sides:

> [W]e bought all of the land. All of his title to all of his property and leased back to him for a homesite some of the property north of the water reservoir.... Bob never really wanted to leave the property. This leaseback arrangement allowed him to still go there. He moved his little cabin back up the canyon.... [T]hey sold the property and get the financial benefit of a royalty.... The deal didn't differ much from the deal we had with Wilder. It was a lease. We were entitled to explore. We were entitled to go on and mine...and we would pay a royalty. (Goldstein interview, KMD vol. 3, 182)

SO YOU DON'T DO A LOT OF WORK FOR NOTHING

William Casburn, who was both a lawyer and a geologist, was hired to begin negotiations with Wilder, who continued to operate his mercury mine, and wanted to be sure he was getting something more valuable in return. Wilder recalled the negotiations that began when Gustafson returned:

> So he came back and then next thing I know,...he came down with Ken Jones. They said, "Gee, we'd like to buy the place. You want to sell it?"
>
> "Nope, not really. We're, you know, we're not interested."
>
> "Well, how about leasing it?"
>
> "No, I don't think we really want to lease it. We're busy, you know, right now."
>
> ...I said, "Just stay out of our area. We're busy. We don't have any intentions of leasing to you. We don't have any intentions of selling to you. So you understand that you don't do a lot of work for nothing and then say, 'Well, you didn't sell it to us,' because," I said, "I don't have any intentions of selling it to you."
>
> ...Finally, Ken Jones is coming down about, jeez, three days a week, I think, and bringing me a bag of donuts and the newspaper. He's coming all the way from Reno and I'm feeling bad. I said, "God, Ken, you've got to be leaving home at four o'clock in the morning to get here. And you're making me feel terrible about this. I don't want to be obligated to you. You're a real nice guy and everything but I

have no intentions of this." It was really kind of disheartening to him because he was kind of crushed about it because he is a nice guy and a real intelligent geologist, not just some half-baked lunatic. (Wilder 1996, 57)

Wilder already knew here was gold at the Manhattan:

I had found gold before. This was not some secret. I did tell them that, too. If I tell the truth on these things and tell everything I know right up front, there's never any subterfuge involved.... Told them I had a gold assay. It was done by reliable people. That was Martin Quist who had Metallurgical Labs in San Francisco.... Mel Stinson was the state geologist. I used to see Mel all the time. We'd either go get lunch or a cup of coffee or something down in San Francisco when I've been out to deliver some mercury, see. And Mel told me, "You ought to get an assay on that because you could have some gold or silver in there."

I kind of went, "Oh, Mel, this is like looking for elephants in the Mojave desert, you know, in the wrong country."

He said, "No, Bill, that thing looks like it." I hadn't paid much attention to it. But I went up to Metallurgical Labs and left a sample. God, I got a call from them through Mike Fopp.... He came up. There was no phone at the mine. And he said, "God, I got a call from Metallurgical Labs. They'd like to talk to you."

So I went and phoned them. They said, "Hey, you ought to come down. Next time you come down, stop in because this is interesting. You've got something there."

I said, "Well, what is it?"

"Well, it's about two thirds of an ounce, .675."

I said, "Wow, that's pretty good." But gold was thirty-five dollars [an] ounce so now we got twenty dollars a ton in gold. We've got mercury that is running thirty, forty dollars a ton. What are we going to be fooling around with gold at half the price? So, I said, "Okay, that's good." I stopped in and talked to them. And I said, "Gee, that's nice." No interest in it though because what good is it? But then gold changed, see, in '73. It was still thirty-five dollars, I think it was. Along in there, someplace, '73, '74 it started going up. Then it went up and went back and up and down. Didn't worry about it. (Wilder 1996, 58–59)

More visitors were coming; one was a lawyer. Wilder began to worry.

> We were busy on mercury and so Ken Jones came down and I didn't hear from him for a long time on that. It must have been, oh, maybe three weeks. Then all of a sudden, I came home from down the plant up to the trailer. And uh, oh, there's a car parked out in front. That's Ken's car and another car, too...it was two cars. I don't know what's going on.
>
> Then Ken says, "Hi, Bill." Real friendly and everything. He said, "This is Bill Casburn. He's a lawyer. He's an attorney with Homestake."
>
> And I went, "Oh, my God. Now what the heck is going on? It's something." I look at big companies like—look out. I'm a very little company and they're a very big company. Look out because they'll step on you and not even know it. I just wouldn't hardly given them the right time, you know. I mean, Bill was trying to be friendly.
>
> I'm going, "Hmm. Wonder what he means by that?" I'm very gun-shy because...I'm listening to everything and analyzing everything as fast as I can, filing it so I can hash it over later. That must have been on a Friday; they stayed for a couple of hours. I think they may have had a bite to eat and talked and then they left. It was a good kind of a thing, you know, funny thing.
>
> I think it was Sunday, Kay and I were at home and all of a sudden, a car drives up. We didn't lock the gate. A car drives up and I'll be darned. I don't recognize the car. It was Bill Casburn and he brought his wife along, Jean.... Then he comes up and he goes, "Hi, Bill."
>
> I said, jeez, you know, I don't know him that well. I'm pretty nervous.
>
> "I brought my wife along. I thought, maybe we could go to town and get dinner. I would like to talk to you some more, Bill, get to know you a little." (Wilder 1996, 59)

Wilders were always hospitable to visitors.

> So we went to town. I said, "Sure, that's a good offer, you know." We went to town and had a bite to eat and talked and came back home. It's a long ways from there, so they stayed over. I said, "Yes, stay over." We have a spare bedroom; they stayed in the guest room. And that was kind of good. Then they left the next morning. It must have been Monday morning, Bill left.

Then didn't see anything for a long time, for a week or something, maybe. Then he came back up and he brought some beer and stuff and it was about five o'clock and getting there, it was dark. It must have been during the winter time because he brought beer up for the crew. That was Billy and all the guys and we had a big bull session, kind of, down at the mill and that. Bill hung around and talked by himself. He'd talk about things. He got to know everybody, kids all know Bill real well and he just got to know how the operation functioned and what was going on all the time and telling war stories, you know, type thing. And it was good. Bill's a pretty good hand at that. He's a good hand. They kept doing this and finally Bill said, "God, well, Bill, we'd like to explore this. We'd like to get a lease before we do any exploration."

I said, "Bill, I have no intentions of leasing at all." "Why not?" And I said, "Well, we're just a little one-horse outfit and we'll get eaten up in this thing. You know, we don't have any legal staff. We don't have anything. We're damn lucky to be functioning, you know."

So he said, "Well, let me bring up a lease next time." He brought up a lease and he said, "Here, look at this. This is our lease." (Wilder 1996, 60)

THIS IS TRADING A SARDINE FOR A SALMON

As Wilder read the lease Casburn showed him, he became more and more wary.

I looked at it. I read about the first two paragraphs and said, "Bill, I'd have to have a wooden head to sign this thing. This thing is so you've got everything. This is trading a sardine for a salmon. That's what it is. You're going to give me the sardine, see."

And so he laughed and he said, "Well, let me see. Let me get some other leases together, here." So he came up and had about five different leases.

And I'd read them and I wouldn't get into two paragraphs and I'd say, "Oh, no, no. You've got this, you've got—hey, we don't have anything. We're giving the whole farm away." So I said, "No, I think we better stay with that."

He said, "Okay, let's take the least restrictive one. Let's go through it. Okay, the first paragraph. What's wrong with that?"

"Well, there's nothing wrong with the first paragraph."

"All right, the next paragraph."

"Well, I mean, this is where you have the right to stockpile any place on the thing, material, and I don't have the right to do anything here, see."

"Well, let's change that, that we have the right to stockpile at your direction. We can do it and you have the right to use any place you want on here."

"Okay, that sounds more like it."

We go on and on and on in terms and then: water. "Well, Homestake has the right to use the water as necessary for drilling."

"Wait a minute. This is great. But what about the One Shot? The One Shot needs water to keep the plant running, keep the people alive out here."

"Well, then, let's work it so that One Shot has the first shot at all the water and any surplus water that's available is available to Homestake." And he did it. (Wilder 1996, 61)

Casburn changed his approach, and Wilder began to soften.

He was good. I'll tell you, I learned a bunch because the way he did this was to say, "Okay, if all the water belongs to One Shot, Homestake has no water rights or no right to any water whatsoever."

I go, "Now, wait a minute. That's not fair either." So, see, he did a little strategy that I've never seen anybody do before. He put me on his side and he's taking my side, now, see. You know, and this all of a sudden, inherently, you've got to know your person. If you've got a guy who's a thief, he's going to say, "Good," and take it. But if you've got a guy that's halfway, you know, that's got some compassion, he's going to say, "Wait, I don't want to take everything from you. That isn't right, either."

"Okay, let's work something out where we split the water."

"Okay, let's do that." And so that's how he would write it up.

We did this for a month and he came up with this lease. And he said, "Well, now what's wrong with this lease?"

...That's how we did this by going [over] every point. I said, "Wait, I've got to look this over more, Bill." (Wilder 1996, 62)

Wilder spent a week alone at a remote place on the northern Mendocino coast, going over the agreement.

I went out to Fort Bragg up the Wage's Creek up there and sat on the beach for a week, brought the camper. You know, we've got a travel trailer. Sat there and all I did was read. The intent of this paragraph is this. This sentence means this and this. Took this thing apart, right down the line. Anything that I saw that I didn't think was right, I put a question mark and then noted it on paper. Asked him and he'd say, "No, they will change it to where it means this."

And he said, "Well, what's wrong with this lease now?"

"Jeez, I guess there's nothing." (Wilder 1996, 62)

Casburn also recalled the negotiations, which took more than six months:

I was under extreme pressure to close the deal because Ken Jones and Don Gustafson were talking about odds of ten to one that Bill's property contained a major gold deposit.... We spent many enjoyable weekends together and I always enjoyed our meetings because he is such an interesting person and a man of many talents.

We had submitted to Bill our standard lease agreement, which with a person like Bill is simply a starting point or basis for discussion. Of course Bill virtually tore the agreement apart.... [E]very change Bill proposed, I asked him to write it in between the lines or I would use Bill's language and write it in and then have Bill initial it, if he agreed to the change. In this way we virtually rewrote the whole agreement as a series of scrawls and marginal notes.... I had long since given up but since our meetings were always very pleasant I hung in there.... In any event...he finally signed the agreement.

When I submitted the signed agreement for review the chief counsel was horrified. He explained to me that when he had said that I could make changes in an agreement he did not mean that I could rewrite the entire agreement. I explained that this was the only agreement we would get, and after review by Harry Conger and Jim Anderson that turned out to be the official version that went into the files. (Wilder 1996, xiii)

Wilder was satisfied that the agreement was fair to both sides:

So we did sign a lease then. And you know what I charged them for a fee for exploration while they had the property, a thousand acres, I charged them $1300 a month. And we supplied them with power, water and...we made the drill sites there for them. We did repairs for them for Longyear.... I felt good about this. We did a lot of work on that exploration.... It was a good deal for them. I knew that.

Mike Fopp told me, "Bill, if they offer you a million dollars, you grab it and run. Don't even pick up the equipment. Get out of here."

I said, "Mike, that's insane. That's not true. I've got my whole life in this thing." And the mine is better than that. It deserves a better shake than that. And it did. I was right. I know that. So it worked out pretty good. (Wilder 1996, 63)

Now Homestake had the land, and they could announce the new project. Harry Conger, the president of Homestake, recalled the big event:

> We finally got the land position in August, 1980.... We made the announcement at three o'clock, when the market closed in New York.... Like we do all our press releases, just standard press releases, only this was no standard release.
>
> [W]e announced that we had discovered a major gold deposit here in California. A couple of TV stations came over that afternoon, and I gave an interview in the board room. Radio stations were calling, and the Wall Street Journal and everybody....
>
> And would you believe...that was the night...that this bozo blew up the Harrah's Inn over in Reno with a bomb in the lobby?... [T]hat night on all the news, instead of our gold discovery, was this damn thing over in Reno.
>
> [I]t was a poor news day for us. But in any case, the fun was the next day, when our stock couldn't open...there were a lot of buy orders for the stock, and there wasn't anybody that wanted to sell. (Conger interview, KMD vol. 2, 180–81)

This was a high point in the history of Homestake Mining Company. They were going to develop a new gold mine and success seemed assured.

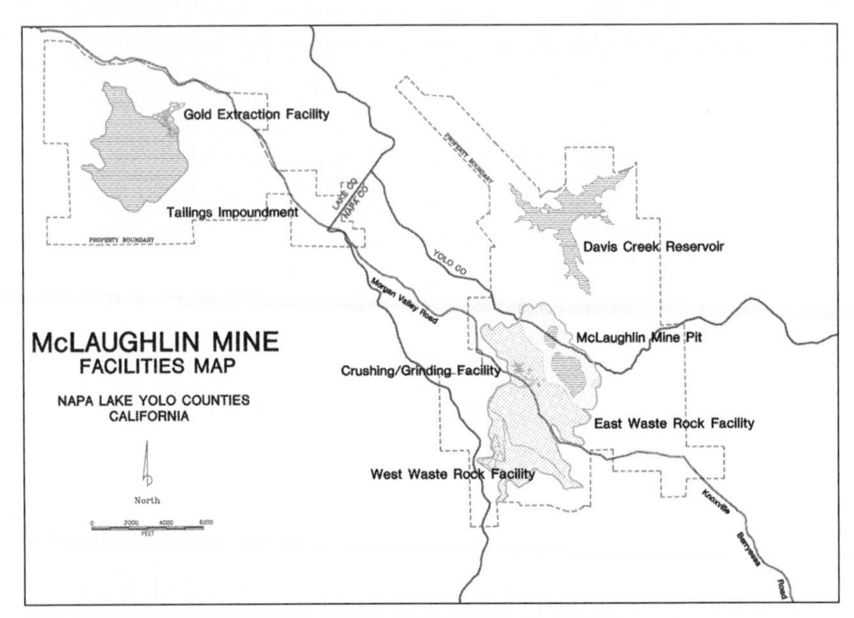

Homestake McLaughlin Mine. (Map courtesy Dean Enderlin.)

Left to right: Rex Guinivere, John Ransone, Klaus Thiel, and Jack Thompson at mine opening reception. (Photo by Manley-Prim. Courtesy Oral History Center, The Bancroft Library, University of California, Berkeley.)

Metallurgical engineer monitoring innovative autoclaves processing control system, 1988. (Photo by Manley-Prim. Courtesy Oral History Center, The Bancroft Library, University of California, Berkeley.)

Waste disposal reclamation site, ca. 1992. (Photo courtesy Jack Thompson.)

Mercury retort, heating recovery to produce liquid mercury, 1981. (Photo courtesy Jack Thompson.)

Underground tunnel, official opening to mine rock samples, 1981. (Photo courtesy Jack Thompson.)

Inception of reclamation while still processing ore, 1994. (Photo courtesy Jack Thompson.)

Aerial view of the original mill plant, 1990. (Photo by Manley-Prim. Courtesy Oral History Center, The Bancroft Library, University of California, Berkeley.)

James William “Bill” Wilder, mine opening ceremony, 1985. (Courtesy McLaughlin Mine Collection.)

Joseph Strapko, geologist and mineral explorer. (Courtesy Oral History Center, The Bancroft Library, University of California, Berkeley.)

Creek aquatic sampling to check on mining residues and status of insects and fish, 1988. (Photo by Manley-Prim. Courtesy Oral History Center, The Bancroft Library, University of California, Berkeley.)

Aerial view of McLaughlin reclamation site, 1993. (Courtesy Oral History Center, The Bancroft Library, University of California, Berkeley.)

Aerial view of reclamation site after seeding, 1992. (Photo courtesy Jack Thompson.)

Autoclaves ore-mining process site, Lake County, 1990. (Photo by Manley-Prim. Courtesy Oral History Center, The Bancroft Library, University of California, Berkeley.)

Aerial view of oxide-tank plant, 1991. (Photo by Manley-Prim. Courtesy Oral History Center, The Bancroft Library, University of California, Berkeley.)

Patrick Purtell, 3 millionth oz. gold bar, 1999. (Courtesy Oral History Center, The Bancroft Library, University of California, Berkeley.)

Furnace workers pouring of gold bars in protective gear, April 29, 1985. (Photo courtesy Jack Thompson.)

Aerial view of blast pattern, 1992. (Photo courtesy Jack Thompson.)

Left to right: Jack Thompson and Sylvia McLaughlin, mine opening ceremony, 1985. (Photo by Manley-Prim. Courtesy Oral History Center, The Bancroft Library, University of California, Berkeley.)

Cat® Hydraulic Shovel loading ore in an open pit, ca. 1986. (Photo courtesy Jack Thompson.)

Left to right: Susan Harrison, Patrick Purtell, Raymond Krauss, Dean Enderlin, and Eleanor Swent at McLaughlin Mine, 2002. (Courtesy Oral History Center, The Bancroft Library, University of California, Berkeley.)

PART 3

CHAPTER 10

PROTECTING THE ENVIRONMENT

Nobody gave a damn in those days;
the environmentalists hadn't yet been invented.

—Philip Read Bradley Jr.

Philip Bradley, for many years the president of the California Mining Association, remembered gold mining in Alaska in the early twentieth century; now Homestake Mining Company was trying to develop a mine seventy-five years later, in California, and the situation was very different. Geologists had found the gold, and the public announcement was made in August 1980. The company still had to obtain permission to operate a mine; in order to get that permission, the neighbors and their representatives in multiple government agencies had to be convinced that their environment would be protected. Otherwise, the gold would have to stay in the ground.

The first Earth Day was celebrated on April 22, 1970, by more than 20 million Americans, and made San Francisco a center of the environmental movement that spurred legislation at all levels. President Richard Nixon, in response to public concerns, created a council to consider government programs; they recommended establishment of a single agency to consolidate research, monitoring, and enforcement. The Environmental Protection Agency (EPA) was established on December 4, 1970, with the mission to protect human health by safeguarding the air we breathe, water we drink, and land on which we live.

The Clean Air Act of 1970 authorized the EPA to establish National Ambient Air Quality Standards to protect public health and public welfare and to regulate emissions of hazardous air pollutants.

On October 18, 1972, Congress passed the federal Water Pollution Control Act, commonly known as the Clean Water Act, with the purpose of preventing pollution, providing assistance to publicly owned wastewater treatment facilities, and maintaining the integrity of wetlands, and on September 30, 1976, Congress passed the Resource Conservation and Recovery Act, which gave the EPA authority to control hazardous waste, including the generation, transportation, treatment, storage, and disposal of harmful materials.

In California, the Environmental Quality Act was passed in 1970. The Surface Mining and Reclamation Act (Public Resources Code, Sections 2710–2796) was enacted in 1975.

On December 11, 1980, President Jimmy Carter signed into law the Comprehensive Environmental Response, Compensation, and Liability Act of 1980, which established the Superfund program to pay to cleanup sites that had been contaminated with hazardous substances and pollutants.

The Sulphur Bank Mine on the shore of Clear Lake, which had operated intermittently since 1865, closed in 1957. It was declared a Superfund site when EPA tests found mercury in the water and the fish.

YOU CAN'T IGNORE THE PEOPLE SIDE

In 1973 David Crouch went to work for Utah International Mining & Construction Company at their headquarters, 550 California Street, in San Francisco. He reported directly to the president, Alexander Wilson, and his initial task was to draft the first statement of environmental policy for the company. As he recalled, "environmental quality was a totally new notion" (Crouch interview, KMD vol. 3, 17).

Crouch had a family background in both mining and health. His father was a mining engineer/geologist who operated a small underground bauxite mine in Arkansas. On his mother's side of the family, all of the men, and some of the women, were either medical doctors or in public health positions. He graduated from the US Naval Academy in 1955 and was contracting officer for construction and nonpersonal services at an air base. After leaving the Navy, he worked for a life insurance company and became a regional manager but found it unfulfilling. He enrolled in the new graduate program in environmental health science at UC Berkeley, and received his master's degree in public health in 1973. Then he was hired by Utah International.

In 1979 the Homestake Mining Company, a block up the hill at 650 California Street, was planning a new gold mine in Napa County. Langan Swent, vice president for engineering, had managed Homestake's uranium mines in New Mexico, and knew that personal and political issues were as important as technical ones; the company had remained nonunion in a region dominated by larger, strongly unionized operations, and won national safety awards. After a search, he recruited David Crouch to be Corporate Manager for Environmental Affairs. Crouch was hired on January 1, 1980, and Swent recalled,

> [W]e set up in the San Francisco office.... [This] gave the person in that job more authority and clout through the company.... He and I wrote a policy...that was a real transition. (Langan Swent 1996, vol. 2, 650, 870)

That policy statement was to govern all company operations. From then on, environmental quality concerns were built into every Homestake project.

Crouch learned that at Homestake the relevant activities were handled in multiple places by the exploration group, "people who were skilled at chipping at rocks and looking at core and directing drilling programs, who had to go down and sit in the local government office and take care of paperwork and try to keep the federal land managers happy or the state people happy" (Crouch interview, KMD vol. 3, 35). He met with James Anderson, vice president for exploration, who was glad to turn over the environmental responsibilities. Crouch then hired Allen Cox, formerly with Utah International, to work in the Reno office, as regional manager for environmental affairs. His assignment was

> to work with the field geologists, to secure the permits for them so they could do their drilling and exploration work, and to accomplish the reclamation that was necessary at sites where we had land disturbance, and generally to make sure that we were doing what we should be doing, and establish good working relationships with the different federal and state government agencies. (Crouch interview, KMD vol. 3, 35)

Crouch also brought Jerry Danni, already on the Homestake staff at Pitch, Colorado, to be the regional manager in the Golden, Colorado, office.

The first core hole at the Manhattan project was drilled in September 1979, and the core sample was driven in a pickup truck to the Hunter Assay Laboratory in Reno. Eventually they had drilled 406 holes, and by March 1980 they had identified a deposit containing 6 million tons with around 0.17 ounces of gold per ton. On October 24, 1979, Crouch made a site visit and began to consider the requirements for environmental impact assessments by both the federal and state regulations. Then he had to estimate how much time and money it would take. By April 1980 he was ready to make his recommendations to the senior management:

> I felt confident that you could permit the project if you covered all of your bases in terms of doing a complete environmental analysis of the existing conditions and if you built into the project planning a recognition of any sensitive areas and...what your requirements might be. (Crouch interview, KMD vol. 3, 43)

The effort began with the exploration geologists, who had to be educated in this new approach:

> Where you have land disturbance, we were required to dress that up and to smooth out places, to backfill pits that you'd use for collecting drilling mud. There were certain guidelines which we needed to assist our field people in understanding about crossing streams and maintaining good housekeeping around the sites, cleaning up after themselves, being careful about oil spills and that sort of thing.... [I]n some respects what we're talking about is just good common sense.... But sometimes people had to be reminded of it because these had not been requirements in past years, and so for some it was a new thing. (Crouch interview, KMD vol. 3, 35–36)

He drew on his experience with Utah International in developing a project with multiple governmental agencies. There he formed an agreement with the federal Bureau of Land Management and an association of local governments. That became the model he used for the McLaughlin project, working with the three counties and the Bureau of Land Management to set up a joint agreement on the required environmental studies.

> And it provided a good forum for everyone to bring their interests to the attention of the rest of the group, and we were able to talk through issues and identify what the sensitivities were on various parts of the project.... People are the ones who make decisions. And people are the ones that form opinions and influence how things happen.... [Y]ou have to deal from a good, solid base of science in terms of understanding the physical aspects of whatever the proposal is, but you can't do that and ignore the people side. (Crouch interview, KMD vol. 3, 15, 40)

This would be the guiding policy for Homestake to develop a modern mine in California.

CHAPTER 11

1981—A NEW TEAM

Mining is a unique business.
—William Humphrey

In 1981 important decisions were made for the new team to work on the Manhattan project; the first was when William Humphrey was named Homestake's first executive vice president of operations. Then he made two hiring choices: one, an innovation essential for the development of a modern mine in California; the other, traditional. Raymond Krauss, an environmentalist, and Jack Thompson, a mining engineer, would work together on the new team. Protecting the environment would be as important as mining and processing the ore.

William Albert Humphrey was born in 1927 in Potrerillos, Chile, where his father was smelter superintendent for Anaconda Copper Company. The family later moved to Anaconda operations in New Jersey and Arizona. Bill studied geological engineering at the University of Arizona in Tucson, graduating with distinction in 1950. He began as junior geologist for Anaconda at Cananea, Mexico, and rose to be executive vice president of Cananea Consolidated Copper Company. In 1975 he was hired by Newmont Corporation as vice president of operations, overseeing mines in Canada, Peru, South Africa, and the United States. When he joined Homestake early in 1981, he had formed definite opinions about the mining business:

> I don't think you can run a mining company with anyone but a mining man. Or a metallurgist, or a geologist, or somebody that's part of the business.... [M]ining is a unique business, and it's distinct, and you should know it, and you should know it from the bottom up. So you know how the workmen think, what their aspirations are, their idiosyncracies [*sic*] are, because all the labor that

> I've seen in mining have customs that they just can't break.... When you work underground, it's a home to you. The mine is a home, and you can't treat underground miners the same way you treat people working at a factory, because it just doesn't work. (Humphrey 1994–95, 139, 140)

In 1981, although the price of gold was down, Humphrey recalled, "We were so anxious to get this mine established and operating, because we thought the gold price was going to be between $500 and $600 an ounce. From the size of the ore body and the grade, it looked like it could be a real winner" (Humphrey 1996, 146).

MAYBE I CAN MAKE IT A BETTER PROJECT

Four years earlier, in 1977, Governor Jerry Brown made two appointments to the California State Board of Mining and Geology. The name of this board had recently been changed to add "Geology" since the state's interests were no longer confined solely to mining. One new member was James Anderson, Homestake vice president for exploration; the other was Raymond Krauss, planning director for Sonoma County. The two attended their first meeting together, frequently traveled together to meetings around the state, and formed a friendship, although they had sincere and differing views of natural resources.

Krauss was born in Athens, Ohio, in 1942, the son of a college professor, and he recalls that as a child he "saw the abandoned coal mines of Appalachia, the old slag heaps and acid creeks and disturbed areas throughout the hillsides" (Krauss interview, KMD vol. 8, 187). He graduated from Oberlin College as a biology major, did graduate work at the University of Michigan, taught at a community college in Maryland, and worked for a planning agency in Puerto Rico. His interest expanded into the relationship of environmental science to the political, how the decisions are made, how the money flows. In 1970 he went to San Francisco to work for a foundation, "doing community organizing, writing grant proposals, and dreaming about how we'd make a better world" (Krauss interview, KMD vol. 8, 196). He was one of the many young people inspired by the first Earth Day celebration on April 22, 1970.

In 1971 Krauss was selected from a pool of more than two hundred applicants to be Sonoma County planning director. There he worked with the case of a company that was mining gravel in the Russian River. By 1977, when he was appointed to the California State Board of Mining and Geology, Ray Krauss was skilled at the myriad elements of putting a mine into operation: systemwide environmental analysis and dealing with a mining company, the neighbors, the opponents, the proponents, the local government, the State of California, and the federal government.

In 1980, on one of their trips together, Jim Anderson asked Ray Krauss if he thought it would be possible to permit a mine in Napa County. Ray reacted quickly. "You've got to be kidding. Napa County, right next to San Francisco, the home of all the environmentalists? No, it's not possible, you're wasting your time" (Krauss interview, KMD vol. 8, 205–6). Later, after feeling him out a bit more, Anderson asked Krauss if he would be interested in helping Homestake get the permits for the new mine project. Ray thought about it for some time before he submitted a résumé.

Then he had lunch at the Washington Street Café in Yountville with David Crouch, environmental affairs officer; Donald Gustafson, the geologist still in charge of the operation; and Vice President Robert Reveles, the Homestake attorney in charge of environmental affairs. For an hour and a half, there was questioning on both sides. The Homestake people were glad to know that Krauss had well-established friendly relationships with the members of the Napa County planning department. He was relieved to learn that the Homestake people were well aware that they would be under public scrutiny, and that they intended to comply fully with all the laws and regulations. He had not visited the mine site, but he knew its reputation:

> It was a remote area that had been badly abused historically, and that there were a bunch of marijuana growers and hippies and other low-life hanging out in the woods up there, and it was a strange and remote area. But I also knew it was in the watershed of Lake Berryessa, and that it was subject to the approvals of three counties—at that time Napa and Yolo County, and very shortly thereafter when we started looking around for tailing sites, Lake County. (Krauss interview, KMD vol. 8, 223)

His colleagues in the environmental community believed that working for a mining company was wrong, and tried to discourage him, but he concluded that the mine would certainly be developed, and "if I'm party to it, maybe I can make it a better project than it would otherwise be" (Krauss interview, KMD vol. 8, 223).

Before making his final decision, Krauss met with William Humphrey, Homestake's new executive vice president of operations. They played a role reversal: Krauss emphasizing that he was practical and would consider cost, Humphrey that he was environmentally sensitive. On June 19, 1981, Raymond Krauss resigned from the California State Board of Mining and Geology and became environmental manager for the McLaughlin Mine, according to his job description, "with responsibility to obtain the permits for the mine and to do the environmental planning for the mine, and to interface with all of the environmental aspects of the mine with the metallurgists and the engineers and contractors, and make sure the requirements of the permits were implemented" (Krauss interview, KMD vol. 8, 225). Humphrey later said, "He turned out to be just a crackerjack. Ray knew all of the ropes about how to handle these different committees, these different state and county committees, and municipal committees, and what procedures to follow so that they felt comfortable about it. I don't know what we would have done without him" (Humphrey 1996, 153).

For the next two and a half years, Krauss worked out of a rented office on Willow Lane in Napa, reporting at first to Donald Gustafson, still project manager. Irma Taylor, an accountant and now a Napa resident, had worked for Homestake before, in Grants, New Mexico. She was hired as the bookkeeper; one of her jobs was receiving assay data from the drillers who were already taking core samples to delineate the ore body. Linda Thomason, also a local resident, was the secretary. Jim Anderson was there, reviewing the drill results and becoming increasingly excited; the others in the office began to call him Sunny Jim.

David Crouch welcomed Ray Krauss and believed that his hiring was one of the best decisions made for the project. "He brought a lot of enthusiasm, fresh thinking, creative thinking, and a real willingness to try to work with people and to make things happen in the right way" (Crouch interview, KMD vol. 3, 53).

"WHAT EVERY MINING ENGINEER DREAMS ABOUT."

Then it was time to transfer responsibility for the Manhattan project from the geologists to the engineers, and Bill Humphrey recalled earlier experience with this critical step:

> It gets very tricky.... [The] geologists cling to a deposit.... It's their baby, and they don't want to let it go.... [They say,] "when we have it a little more well defined and the process a little more developed, we'll turn it over to the engineers so they can operate it." (Humphrey 1994–95, 148, 149)

A project manager at this point needed more than just technical skills, and Humphrey remembered a personable young engineer who had worked with him when they both were with Newmont Mining Company. Jack Edward Thompson Jr. was born in Cuba, where his father, a graduate of the Colorado School of Mines, worked for a geophysical company. Jack Jr. graduated from the University of Arizona as a mining engineer; during vacations he worked as draftsman and surveyor for the highway department and underground as a laborer at the San Manuel Mine. More recently, he managed the open-pit Dawn Mine in Washington. When Humphrey called to offer him the job of manager of the McLaughlin project, he jumped at the chance:

> It was an opportunity to be involved with a project from the ground up, where all we had was the drill data and we had to calculate the ore reserves, design the mine, construct it, and then run it. Really, that's what every mining engineer dreams about. (Thompson interview, KMD vol. 7, 32)

On August 1, 1981, Jack Thompson was hired and moved with his family, first to a rented home in Napa, and later to one they bought in The Rivieras, on the flanks of Mount Konocti in Lake County. He was reporting then to Jim Anderson in the exploration group; the geologists had drill rigs at the Manhattan site, and had identified an ore reserve of 20 million tons, valued at 0.16 ounces of gold per ton. Jack Thompson's first task was the feasibility study, as he recalled:

> An engineering and financial study that analyzes this deposit, proposes a development scheme, comes up with the capital and operating cost estimates, and does the financial analysis to say whether it pays for us to invest this money or not.
>
> Now, my group, as operators and engineers, look at the data that the geologists handed us a little differently because our needs are different.
>
> ...We redid the drawings and the sections and the plans so that we could use them for mining...and planning of the pit...we had a stack of sensitivities, of computer printouts, you wouldn't believe.... We used a lot of reams of paper before it was all done.... We looked at all the alternatives.... What happens if recovery is only 80 percent instead of 90 percent...what happens if costs are more...What

> happens if the project gets delayed, what does that do to the rate of return.
>
> ...[In] 1982, it [the price of gold] hit a low of $280.... We spent a lot of time trying to decide what prices to use and in the end...[w]e had a base case and then we had a high and a low.... [W]e then took [it] to the directors and said, "Here is our analysis of this. We'd like to spend a couple of hundred million bucks here."...[A]nd they gave us the go-ahead...in '82.... [I]t was a very exciting time, let me tell you. (Thompson interview, KMD vol. 7, 39, 40, 41, 42)

He hired a former schoolmate, Mike Attaway, to do the mine design. He brought in other Homestake employees: one was Knollie Sell, an accountant from the uranium mines in Grants, New Mexico, whose father had also worked for Homestake. From the uranium property at Pitch, Colorado, Thompson brought Tim Janke, who became one of the key engineers on the open-pit design for the new mine. A husband-and-wife team, Roger and Pam Lucas, also came from Pitch; Roger became a supervisor in the pit, and Pam worked in human resources.

Jack Thompson and Ray Krauss liked each other from the start, and Thompson believed that they grew and learned from each other:

> [Ray] learned quite a bit, I think, about how business works...but he's also had a moderating influence on us in terms of advancing the state of the art in environmental protection and how do you do business in the nineties now. (Thompson interview, KMD vol. 7, 35)

Looking back later, Ray Krauss described the McLaughlin culture and Jack Thompson's management style:

> There was this sense of teamwork and a sense of mutual respect at all levels.... Jack always was very encouraging to his employees. He kind of set the relationship as, Well, I hired you because I think you're the best person available to do this job. Now go do it. And I'm not going to get in your way. Your real obligation is to keep me informed, tell me before the train comes off the tracks. I don't want to know if you had a problem after it has happened; I want to know before it has happened. But go and solve that problem and manage your responsibility. It was a real flat organization. We had tremendous communication horizontally.
>
> [P]art of the Mclaughlin culture, was to be willing to be open to the public, to be transparent, really, to be in a fishbowl—all of our

> activities, all of our monitoring. Anybody that had any questions, "Come on out and look." "Can I take pictures?" "Sure." Other than getting into the refinery and handling the gold, there was nothing really that we wouldn't share with people. (Krauss interview, KMD vol. 8, 383)

In 1986 economist Simon Strauss published a book with the prescient title, *Trouble in the Third Kingdom: Minerals Industry in Transition*. In it he summarized the trouble: in the three kingdoms—animal, vegetable, and mineral—only minerals are nonrenewable, and "the physical acts involved in mining and processing of minerals disturb some people. Inevitably these acts involve environmental consequences. The earth's surface is changed as soil and rocks are removed to provide access to mineral deposits" (Strauss 1986, 1).

His book ends with a long look forward by Sir Arvi Parbo, Australian mining engineer and chairman of Western Mining Corporation, Alcoa of Australia, and of the Broken Hill Company Pty Limited: "By 2135...there will hardly be any miners underground. Minerals will be won either by robotized machinery or by biotechnology employed in situ to convert minerals into a soluble form.... This is indeed a long look forward" (Strauss 1986, 210).

Now, in 1982 in California, a mining company was proposing to change the earth's surface, using real men and women, not robots or biotechnology, and people in Napa, Lake, and Yolo Counties were concerned over the environmental consequences.

Homestake had a new team in place to develop a modern gold mine. Next, they would need to obtain the permits to mine, and to convince the people there that the mine would not harm their air or water, and that the site would be properly reclaimed later.

CHAPTER 12

ENVIRONMENTAL PLANNING

It was fast track, it was do it, do it now

—Raymond Krauss, environmental manager

Raymond Krauss was hired as environmental manager for the McLaughlin Mine, with this job description: "With responsibility to *obtain the permits* for the mine and to do the *environmental planning* for the mine, and to *interface* with all of the environmental aspects of the mine with the metallurgists and the engineers and contractors, and *make sure* the requirements of the permits were implemented" (Krauss interview, KMD vol. 8, 225; emphasis added). Of the four specific assignments, the first was immediate: to obtain permits for the mine. The second, to do environmental planning, in fact had to be done first, in order to obtain the permits. All the tasks involved interfacing with the staff. To make sure the requirements were implemented was a commitment of more than twenty years.

Krauss and his team faced two challenges in obtaining the permits. One was technical, to evaluate and plan for the economic and environmental impacts of the mine: to calculate amounts of earth and water; numbers of employees, machines, buildings, roads; and then compile the documents required by numerous agencies. The other was social: to meet the people in Napa, Lake, and Yolo Counties who had prejudices and fears, hopes and personal obligations, and to persuade them that this mine would benefit them.

Ray Krauss recalled his first visit to the Manhattan site. Bob Ackwright, project manager, drove them up through the Napa Valley to Angwin and Pope Valley, then on an unpaved road along the Eticuera and Knoxville Creeks, crossing low-water fords, and approaching from Lake Berryessa. Krauss was encouraged:

> [The mine sits] on the ridge between the Putah Creek and Cache Creek watersheds. And...because it is high on the watershed, you don't have a lot of through-flowing water to have to manage from an environmental protection standpoint. It's a fortunate site in that respect. (Krauss interview, KMD vol. 8, 235)

When they came to Knoxville, they saw relics of over a hundred years of production at the mercury mines. Bill Wilder was still processing some ore as well as mercury batteries at the Manhattan, and Krauss, like so many others, was impressed with the collection of surplus equipment, including old buses that were used for workshops and storage.

Homestake had fourteen rigs drilling there, with geology students from California State University at Chico collecting and logging the drill core in a trailer the company set up for them. They were excited about the work, and also found time for fun:

> There were probably twenty-five or thirty geologists all camping in an old hunting shack called the Correa hunting shack down on the western slopes below the Manhattan mine...and they had hired a cook and she would drive into town every day and come back with a pickup truck load of—steaks and beer, I think, was their major sustenance. (Krauss interview, KMD vol. 8, 236)

Wilder had a diesel-powered electrical generator and had strung wires to the shack to power a TV and several refrigerators. Krauss thought the scene was kind of like a submarine:

> I think when you went on to day shift you would roll up your sleeping bag and the graveyard shift would come in and take over your bunk.... [T]he previous owner had made a kind of...swimming hole that had been rock-lined and they had filled it with water. So they would come back after their shift and jump into this swimming pool and drink their beer.... Of course everybody was excited about the core they were seeing, and comparing geologic notes, and trying to make sense of it. (Krauss interview, KMD vol. 8, 236)

The core samples were shipped to the lab in Reno for splitting and assaying. The data came back to the Napa office and Irma Taylor entered it into the computer model. At first the data were sent back to the mine on paper for the geologists to generate the maps that directed the drilling. When the Morgan Valley Road was improved, Homestake installed a microwave link for telephone communication and the information flow improved. Krauss recalled,

> Oh, I thought it was exciting…from an environmental standpoint, because the site had been so dramatically disturbed.… [It] was obviously an opportunity for an environmental improvement.… [It] was remote enough so that people were poaching it continually, and off-road vehicles were using it. True or not, the rumors were that some of the old mercury workings had been occupied by the Zodiac killer[1].… [D]espite the fact that it appeared to be remote and pristine, it had been ecologically totally turned upside down: all of the oak woodlands were cut for fuel for the retorts; all of the grasses were no longer native grasses, but European exotics introduced by grazing; and there was a tremendous amount of erosion as a consequence of overgrazing. (Krauss interview, KMD vol. 8, 238)

Krauss and Homestake attorney Dennis Goldstein had to prepare two crucially important documents for approval by the county supervisors before permit applications could even be considered. The federal EPA required an Environmental Impact Statement. The California Environmental Quality Act required an Environmental Impact Report. The requirements were similar, but not identical; Krauss and Goldstein prepared a single document that met criteria for both.

Before the permit applications could be submitted to the county officials, an independent third party had to review and comment on the plans. Because most of the project then was in Napa County, it was the leading agency in the review process, so Krauss worked with James Hickey, Napa County planning director, to set up an Environmental Data Advisory Committee to satisfy this requirement. From each of the three counties, Napa, Lake, and Yolo, the committee had a supervisor; an elected official; and a planning director, who was appointed. The Bureau of Land Management had two representatives, resource management specialists, who were federal appointees. There was one hired staff person, and a hired Environmental Impact Report consultant. The chairman was Harold Moskowite, the Napa County supervisor from the district where the Manhattan Mine was located.

Meanwhile, the California legislature, concerned that the complicated permitting process interfered with economic development, tried to make it

1 The Zodiac Killer was a serial killer in Northern California in the late 1960s and early 1970s. He signed the name "Zodiac" to letters sent to the local newspapers.

easier by passing an act designed to facilitate the granting of permits. This put even more pressure on the Homestake team, as Krauss explained,

> all of the requirements for a permit had to be laid out up front, and there had to be a determination that an application was complete. Once that determination was made, you couldn't amend it, you couldn't require further studies or additional analyses after you had made that decision, and the permits had to be processed in a set period of time. There were twelve months established as the limit for preparing an environmental report, and eighteen months from beginning to end to issue the permit. (Krauss interview, KMD vol. 8, 212)

Planning for a modern mine included reclamation of the site after mining stopped. A ghost town of crumbling buildings and oozing waste dumps was no longer considered picturesque. At meetings in San Francisco, various ideas were floated for post-mining at the Manhattan, as Krauss recalled:

> Fundamentally, we went through an interesting exercise with the Environmental Data Advisory Committee and Homestake management to select a post-mining land use, which is the starting premise of any reclamation plan.... [W]e talked about—donating the site to the counties for public recreation and parks, and returning it to ranch land, or developing as...housing...[or] Palms Springs–type resort...Golf course and swimming and all. And one of the alternatives was to use it as a research facility for environmental studies for the University of California system. (Krauss interview, KMD vol. 8, 258)

This was the plan that appealed to Krauss:

> Looking at what assets would remain at the site at the conclusion of mining...one of the things that jumps right out is the accumulated scientific data—the baseline data starting with aerial photography and topographic mapping, and mapping of soils, and soil types, and soil depths, and soil capabilities, and site hydrology, and vegetation and wildlife communities and habitats, and baseline water quality, and all of the meteorological data that we had to collect and continue to monitor for our permit compliance requirements. And all of our ongoing monitoring data...years of aquatic ecology data on all of the surrounding streams and on the freshwater reservoir

> and succession data on the plant communities and the reclaimed areas.... [All] that accumulated information, it has a tremendous educational value well beyond any benefit for Homestake, for our requirements to meet our permit requirements, and our monitoring requirements. (Krauss interview, KMD vol. 8, 258)

Bill Humphrey liked the idea of an educational facility, for different reasons, as Krauss recalled:

> If the price of gold went up to 2,000 bucks an ounce, we might well re-mine that site. So Bill was very concerned that we not propose a reclamation use that would preclude future mining. And I can remember him saying, "Boy, if we turn that into a park, all those people that recreate up there are going to refuse us the opportunity to go back and re-mine it. We don't want to do that." (Krauss interview, KMD vol. 8, 259)

There were other reasons to support this plan:

> [T]his idea of an educational facility was a selling point...to the environmental community and to the local community. Certainly Napa County was very supportive of it. I can remember Jim Hickey in a public hearing saying: "These guys are going to leave one big hole in Napa County, and this educational facility is a reasonable compensation for that un-back- filled mine pit."...From our standpoint, it also involved a minimum of additional cost—we did not have to create recreational facilities; we did not to have to install new infrastructure. (Krauss interview, KMD vol. 8, 259)

Yolo County members of the Environmental Data Advisory Committee concurred:

> [T]hey didn't want recreation because they were concerned that creating a recreational site would incur demands for public services that they could not afford, that they would have to improve the quality of the roads to the site and they would have to provide sheriff's patrols, and emergency response for injuries. (Krauss interview, KMD vol. 8, 259–60)

Lake County had its own reason to like the idea, that it would continue to provide some employment for the residents. Krauss could go ahead with the final aspect of environmental planning:

> [The educational facility] very quickly became the preferred alternative. Then...the reclamation plan proceeded to look at the physical management of the site, the availability of soils for capping and closing the waste dumps, and we inventoried all of the soils that we needed to reserve and stockpile.... [O]ne of the innovative aspects of the reclamation plan is the salvage of the soil from the Davis Creek reservoir site. All of that soil was stripped and stockpiled outside the reservoir margin, and is now being used in waste dump closure and mine area reclamation. (Krauss interview, KMD vol. 8, 260)

The pieces of the puzzle were beginning to come together. Krauss and his team assembled the Project Description and Environmental Assessment, not just describing what was proposed, but

> actually documenting the environmental analysis we had gone through and the alternatives that we had considered and rejected, and the environmental rationale that we had used in coming up with this particular project design, and our own assessment of what the environmental consequences of the project would be...those things that were incorporated into the project to mitigate those environmental impacts. We agreed that we would continue to do post-closure monitoring for up to fifty years or until such time that everyone was satisfied that there was no threat to water quality. It was essentially our own...EIR [Environmental Impact Report]/EIS [Environmental Impact Statement]. (Krauss interview, KMD vol. 8, 261)

The agencies welcomed the Project Description and Environmental Assessment that gave them a clear overview of the total project:

> [T]he agencies were quite surprised to see a comprehensive set of documents of all the information relevant to the permits. Historically, people would cut it into pieces, and just give the air quality information to the air quality districts, and just give the water quality information to the water quality districts. (Krauss interview, KMD vol. 8, 261)

Some environmental work was already happening at the Manhattan site. The D'Appolonia consultant firm, based in Irvine, California, was hired, with Ed Sirota as their project manager, to compile a complete inventory of the plants in the region in order to be sure that survival of the plant population, especially rare plants, would not be endangered by the necessary losses. Local botanists, familiar with the area, were subcontracted to work on this survey. Three air quality districts had jurisdiction for managing air quality: Bay Area, Lake County, and Yolo-Solano; stations had already been set up at the mine site to measure air quality.

At offices in Reno and Denver engineers were planning the technical side of the project, selecting a process for the ore, and engineering the facilities. Krauss linked all these efforts, often flying from office to office:

> [A]ll of the pieces of the puzzle were coordinated. As metallurgical decisions were made, those were passed on to the engineering group, and as the engineering group moved forward, they needed information on permitting requirements and restrictions, they'd ask us for those answers.... It was fast track, it was do it, do it now, in the shortest possible period of time.... [I]t put a lot of pressure on everybody.
>
> [I]t was an interactive process. We defined all of the elements of the project and pursued them all simultaneously and met to work together to get them all reconciled. (Krauss interview, KMD vol. 8, 230)

Traditionally, an ore-processing facility is adjacent to the mine, but the final plan here had to be different. The process for recovering the gold used cyanide, and this had a special local connotation, as David Crouch recalled:

> [M]ercury had been there for a long, long time, and people knew about that...it was viewed...[as] more manageable. But cyanide was something that they used down at the [San Quentin] prison, in gas chambers, so it just conjured up a much more emotional view of things. (Crouch interview, KMD vol. 3, 50)

The outflow from the processing plant, called tailings, would contain some cyanide and had to be safely impounded, and the best site was several miles from the mine. If the ore were to be processed at the mine, the tailings would have to be pumped to the containment site. Crouch recalled,

> [I]t was a safer thing to grind up the ore at the mine site and put it in a slurry form and pump it through a pipeline...as opposed to having a similar pipeline five miles long that had the tailings with some cyanide in it.... [T]hat...eliminated a concern about spillage or pipeline breaks.... Lake County is where...the tailings impoundment was, right adjacent to the mill site...roughly some five miles away [from the mine]. (Crouch interview, KMD vol. 3, 49, 51)

The key to George Hearst's success, according to Homestake lore, was that from his earliest days in Gold Run Gulch in the Black Hills, he always secured water rights first. If the metal was there, precious gold or silver, or base copper, lead, or zinc, water for processing was essential in order to recover it. Krauss recalled,

> One of the smartest things that was done, and . . . it was Langan [Swent] and Dave Crouch, was to make a very early filing for water rights. The filing for water rights preceded all of our other permit applications. It had been advertised for public notice long before there was much public awareness of Homestake or the McLaughlin Mine, and there was no opposition to our proposal for water rights. (Krauss interview, KMD vol. 8, 230)

The plans were made, extending the project into three counties. The project spanned a thirty-mile area, from the tailings pond and processing plant in Lake County, past the central mine pit in Napa County, and beyond to the Davis Creek impoundment in Yolo County.

Now Homestake had to obtain the other permits to develop the mine.

CHAPTER 13

INFORMING THE PUBLIC

Care and feeding of the network.

—Raymond Krauss, environmental manager

Some of the groundwork for permit applications was done before Ray Krauss was hired. In 1980, before the announcement had been made to the general public, geologist Donald Gustafson, attorney Robert Reveles, and environmental planner David Crouch advised the county boards of supervisors that Homestake wanted to develop a gold mine. They did the same again before the second announcement in 1981, letting them know in advance, so that they would not be surprised, and could ask any questions.

Homestake hired a public relations firm to help educate people who were unaware of modern mining practice, to assure them that Homestake would not endanger the air or water and would clean up after the mine closed. Some of their costly efforts were actually counterproductive. This was vividly illustrated at the town meeting in Guinda, where Ray Krauss assembled a group of local residents to learn about the mine plans; the meeting was interrupted by a helicopter bringing in a TV crew. The result was a public relations disaster.

Staff members on the spot were more helpful. Jack Thompson became project manager when it transferred from exploration to operation; he moved with his family to Lake County, and they became members of the community. They enrolled their children in the local school, and Linda Thompson's volunteer service there was welcomed. Ray Krauss was already well known there. Along with attorneys Reveles and Goldstein, and geologist Gustafson, they were effective in the grassroots work of scheduling public information sessions, attending supervisors' meetings, and going to community meetings.

Ray Krauss summed up his rationale:

> These local permitting decisions...are ultimately political decisions, and one can't expect a member of the board of supervisors to support our project...if...it would cost them their seat, if they would be unelected to the office at the next election. So obviously the only way to ensure a positive decision was to demonstrate to them that a majority of their constituents were in support of our project. (Krauss interview, KMD vol. 8, 346)

It was essential to inform the public about the plans for the mine. He called it "care and feeding of the network," and made a long list of organizations and individuals in the three counties. Then he showed it to Jack Thompson, who at first said, "Well, this is impossible! There's no way we can deal with all of these people" (Krauss interview, KMD vol. 8, 345–48). Krauss replied,

> "They're all going to have a say before the time is over"—by the time you add up the media and the environmental community and the property owners and the neighbors and the regulatory community, in each of three counties. It was a formidable list. And ultimately we did communicate in one form or another with all of them.... [A]nyone who inquired about the project or anyone who was a potential vendor to the project or a neighbor or all of the political folks, any of the decision makers, anybody that might have an interest in the project. (Krauss interview, KMD vol. 8, 345–48)

He developed a newsletter with a final mailing list of over seven thousand people in the three counties.

Jack Thompson talked to local groups:

> I probably easily gave thirty different talks in those couple of years—Sacramento, West Sacramento, Davis, Woodland, places like Esparto, Williams.... Of course, all of the Lake County towns: Kelseyville, Lakeport, Napa, Calistoga—any place where there's anybody who felt that they were going to be impacted, we were there talking to them. (Thompson interview, KMD vol. 7, 68)

He recalled a beneficial arrangement made with Sonoma State University:

> We had some archaeological sites...over in Yolo County, right on the shore of our proposed reservoir. Ray Krauss used his connections over at Sonoma State and got the professor to orchestrate it so that

> his summer classes were at this location, and they excavated it all for us, and for nothing really. We gave them a trailer, an old used trailer, and put a power line down there, and brought them water from time to time, but they'd be there every summer working on archaeology...instead of becoming a problem, it was a resource for the local community. (Thompson interview, KMD vol. 7, 112)

John Drummond, attorney for Lake County Schools, recalled Thompson's effectiveness at winning friends in the community:

> Jack Thompson and his wife...were the most wonderful people that you would ever meet. Linda was a parent volunteer and she helped at the school; she baked cookies; she did everything. Both Linda and Jack were members of community groups and there wasn't a single community activity that they didn't support. They genuinely cared about everybody; it wasn't phoney. (Drummond interview, KMD vol. 2, 274)

YES, WE GOT AN EYEFUL

John Ceteras, the owner of Blue Heron Farm in the Capay Valley in Yolo County, recalled a tour led by Don Gustafson, the geologist:

> [B]efore anything happened up there...Gustafson took us in his four-wheel drive...about four of us.... We got to see the old mercury mines, the old equipment and abandoned containers, God knows what was in them, piled in the arroyos and the ravines, and all the good little waterways that were obviously polluted from misuse over the years. Yes, we got an eyeful.... The main pit area where they were looking, where they were finding their micro-amounts of gold, was pretty well chewed up already from the mercury miners.... He talked about the fact that if Homestake's plans came to fruition, they'd be cleaning out all...the junk...that was up there, and...that Homestake would be complying with all the rules and regs that were now in place. (Ceteras interview, KMD vol. 2, 79)

A PARTICULAR EFFORT

The Sierra Club merited special attention, as Krauss recalled:

> [W]e] made a particular effort to talk to them because they were in a position to be effective adversaries.... [T]here were two women

> who were members of what was called the Conservation Committee of the Yolo-Solano Chapter of the Sierra Club. And I met them at an early information meeting and invited them to the site, and we spent a whole day touring the site and talking about the environmental issues related to the project and the role of mining in our society.... [T]hey were very skeptical about the point of view that I represented.... But I promised them that we would provide them copies of all of our environmental studies, and we did that. Every time we did a series of studies on potential tailing pond sites or permeability studies on the underlying geology or rare plant studies or studies of faulting in the area, everything that we did, a copy went over to the Sierra Club. (Krauss interview, KMD vol. 8, 348)

The Project Description and Environmental Assessment played an important part in the information effort, as Krauss recalled:

> You got four big volumes—the project description and the environmental assessment—that included the reclamation plan, a summary of the environmental data, and the environmental impact analysis... And then a three-volume compilation of all our environmental baseline data, beginning with soils and geology, and geochemistry of the waste...and the ore, and geochemistry of the tailings.... That's the backbone of the educational facility concept.
>
> ...[W]e had whole truckloads of them—I spent a long time hauling them around to each of the agencies, and then we put them in all of the local libraries for the public to review; and we provided copies to...anybody that had interest.... Gradually, over time, we began to be acknowledged as having a presence in the community. (Krauss interview, KMD vol. 8, 261)

A SUBSTANTIAL SETBACK

Although it was not required by law, Krauss arranged to hold public hearings in all three counties, not just in Napa County. Napa County supervisors saw the mine, which was remote from the commercial center, as a windfall in tax revenue. Lake County felt much more impact and it was generally positive: extending power lines, improving roads, creating more jobs and housing. Spirited opposition came where it was least expected: Yolo County.

Krauss scheduled the first meeting at a place convenient for most residents, the Grange Hall in Guinda. It did not go well, as he recalled:

> It turned out to be a substantial setback...one of those circumstances where not all of the elements were under control...the exploration group had previously retained Hill & Knowlton, the San Francisco public relations firm, to film periodically the activities involved in the discovery and development of McLaughlin. Without telling me, they had decided, since they were going to be filming from a helicopter aerial views of the McLaughlin mine site, that they might as well take advantage of the proximity and just pop over the hill and land their helicopter next to the grange hall and unload their equipment and capture some footage of the first public hearing in Yolo County.
>
> Well, this was very much a surprise to the locals as well. All of a sudden, instead of being Jack Thompson and Ray Krauss there in our shirt sleeves discussing with the planning commission their technical questions about what we were going to do and how it was going to work, we had a public presence that we didn't anticipate at all. The helicopter landed, and the kids were all running around, excited. This was the biggest event in Guinda since the end of World War II or whatever.... Of course, it just put a totally different perspective on who Homestake was. All of a sudden we were a bunch of city slickers with this big public relations firm who were coming into their little community to manipulate the outcome of a public process.
>
> ...Here you are, in this little country grange hall, and the next thing you know they've got...film cameras and television lights and soundmen and microphones set up everywhere. I mean, it totally disrupted the flow of the hearing.... It was an obstacle that we were going to have to cope with. (Krauss interview, KMD vol. 8, 287, 288, 289, 426)

The Homestake team then opened an office in a historic building near the courthouse in Woodland and set up the topographic model of the area that showed the mine, the mill, and the tailings pond. They put up bookshelves to hold all the technical studies, maps, and drawings. They hired a receptionist, Katie Knisley, who had been the secretary to the county sheriff. Krauss recalled,

> She was a wonderful, wonderful woman. She knew everybody who ever lived in Yolo County, and they knew her. So she staffed the

> office, and anyone that had any questions, if she couldn't answer them, she would put them in touch with one of us. (Krauss interview, KMD vol. 8, 345)

The building had marble floors, big staircases, and open courtyards; it was an ideal space for a big reception. They hired a mariachi band and a caterer and invited everyone they knew in Yolo County, as Krauss recalled:

> But we threw open the doors, and people came. Over the course of the evening, all five of the members of the board of supervisors showed up. It was real clear to them that there was a substantial constituency in support of the project, and it was clear to the community and it was clear to the press, which was the demonstration that needed to be made. (Krauss interview, KMD vol. 8, 346)

Ray Krauss, Robert Reveles, Jack Thompson, Dennis Goldstein, Don Gustafson, and others had done all they could to inform the public and win their support. Now it was up to the county supervisors to vote on the permit applications.

CHAPTER 14

OBTAINING 327 PERMITS

It was a long and hard road.

—Dennis Goldstein

Homestake attorney Dennis Goldstein summed up the effort to obtain 327 permits for the Manhattan project:

> Our job was to persuade the Boards of Supervisors and the Planning Directors and the Planning Commissioners that what we wanted to do made sense. It was beneficial to the county. It wouldn't put burdens on the county. It wouldn't bring in unreasonable traffic... unreasonable sewer... [and] water demands and development demands that cost... lots of money.
>
> ...It was a long and hard road...with lots of...studies, lots of hearings, lots of proceedings, lots of questions to be answered, hours and hours and hours and days and months of hard work—a very hard working time. (Goldstein interview, KMD vol. 3, 192–93)

Ray Krauss, who had the primary responsibility to obtain the permits, had a more dramatic recollection:

> I tell people I used to have this terrible dream where there would be this faceless Greek chorus of thousands, and I was trying to make them all nod their heads up and down at the same time.... [That's] what we were trying to accomplish—we needed to get all of these multiple regulators and decision-makers to give us a positive go-ahead on a timely basis. (Krauss interview, KMD vol. 8, 263)

Before each jurisdiction could consider permit applications for the mine, the planning commission of the lead agency had to certify that the reports were properly done. There was no real opposition to naming Napa County the lead agency; most of the project at that point was in Napa

County, and the officials in the other counties were not eager to take the responsibility. Each of the five Napa county supervisors represented a particular district, and they appointed the five planning commissioners who acted for the entire county. The planning commissioners first reviewed the permit applications and, when they agreed, they could present them to the county board of supervisors for final approval or rejection.

David Crouch recalled the complexity of the project:

> The ore body was in Napa and Yolo Counties. And Lake County is where the mill was sited, and the tailings impoundment was constructed, adjacent to the mill site.... [B]ecause we had the involvement of three counties, that just exponentially made the permitting process that much more complex.... Each of the counties had its own air quality district, and...Napa County is a part of the Bay Area Air Quality District, which encompasses some nine counties in and around...San Francisco.... So it was a metropolitan air district... [with] a sophisticated, well-qualified technical staff, but they had had no experience with mining.... In Yolo County you had a very small air department...more concerned with dust...because their experience was agricultural.... In Lake County...the air district had a lot of experience and sometimes contentious experience with the Geysers geothermal generating plants...they were, you might say, sensitized to air issues associated with resource development. (Crouch interview, KMD vol. 3, 51)

The federal Bureau of Land Management and the national Forest Service also had jurisdiction over some of the property.

WE SERVED THE PUBLIC'S INTEREST

In 1981 Jay Corley was appointed to the Napa County planning commission by three supervisors, and after his first meeting was made chairman. Corley was born in Evanston, Illinois, graduated from Stanford University in economics, and earned an MBA degree in strategic planning from Pepperdine University. He started several successful businesses in Southern California, became interested in food and wine, and moved to Napa in 1970 to establish a vineyard. Because he descended from a colonial Virginia family, he felt a personal link to Jefferson's interest in viticulture; he named his vineyard Monticello.

When he took office as chairman of the Napa County planning commission, the county already had a general plan that recognized the distinct constituencies: commercial and agricultural. Homestake people had prepared dozens of documents and had discussed them with the planning staff before presenting them to the Napa planning commission at public meetings that were all recorded. At these meetings, the mine's neighbors voiced their concerns. Corley had not met any of the county supervisors or Homestake mining people except at public hearings; he had never visited the mine site. He was sure that although Homestake subsidized the review process, the company did not in any way control it. He recognized several different challenges:

> [Napa County] has about 100,000 residents, and very few of them live anywhere near the mining property. The property and the land in that area have been mined off and on for many, many years.... [It] was not agricultural land.... It wasn't vineyard land;...it was... mountain land.... [The] longest discussions that we had concerned the process, because this mine was not the old-fashioned mine where you crunch it up.... [We] were talking about huge tonnages of material going through a chemical process and extracting minute amounts of precious metal.
>
> And so the process was for us to fairly and patiently hear a series of speakers discussing their concerns.... They were nervous, they were apprehensive...they just knew that this big project somehow was going to change their life.
>
> ...[It] was our planning staff that did it. It was not the commission.... It was a quality effort, and there was an awful lot of patience.... [The] Homestake people knew that it made a heck of a lot more sense to get everything in line and do it right.
>
> ...[We] served the public's interest by hearing all their dialogue, and concluded that the application was fair and reasonable and proper, and they should go ahead.
>
> ...[N]o one has been damaged that I know of and in fact, we've all benefitted from the commerce of this. (Corley interview, KMD vol. 2, 240, 241, 242, 247–48)

IT WAS A BIG WINDFALL FOR THE COUNTY

Harold Moskowite, Napa County supervisor and chairman of the Environmental Data Advisory Committee, represented the district that encompasses the mine location, and thus he was a key person in approving the permits. He was born ten miles from his home at Moskowite Corners and was a third-generation rancher there; except for the three years he spent in the Navy during World War II, this had always been his home. He remembered when the population of Napa was 3,500, before it grew to about 65,000. After his first ranch was submerged by the Berryessa Dam, he moved up into the hills that enclose Napa Valley to the northeast. The ranch had always run cattle and sheep, and in 1975 he planted 175 acres of vineyard, and built an irrigation dam. He had no inclination toward politics, but in 1975 was urged by a number of people, including one of the supervisors, to represent his district.

Early on, Moskowite went with Don Gustafson to see where they were drilling. He knew the country well, because he used to go there for decorative rock for fireplaces. Then there was a meeting at Oliver's Restaurant in Napa:

> Yes...they outlined what they wanted to do and...they were talking big things and...told us what they were going to do and how they were going to do it...Napa County didn't have a mining ordinance because...Napa County wasn't a mining county. I mean they had some little mines up there in Knoxville, but they didn't have to have permits to do it.... [P]eople didn't even know where Knoxville is.... There's no population up around there.... Forty miles, fifty miles from town—and there was nothing in between.
>
> [T]hat's some of the conditions Napa County set out, that they didn't use Knoxville or Berryessa Road...because it's a substandard road, and Napa County didn't want to bear the expense to have to put in a road.... [T]hey'd come in the Lake County side.... I think they cooperated with the county very well.... [T]hey did everything they said they were going to do, and more.... We tax the gold in the ground, we tax the stuff that's on top of the ground.... We're not providing any services up there.... [I]t was a big windfall for the county. (Moskowite interview, KMD vol. 6, 43, 44, 46, 48, 51, 54)

THEY DIDN'T COME IN LIKE GANGBUSTERS

One of the decision-makers, Walter Wilcox, was Lake County's second-longest serving supervisor on record, serving from 1979 to 1995. He was born in Flint, Michigan, in 1928 and moved with his family to Southern California in 1945. By 1976 he had owned and managed seven corporations and served on the air pollution committee of the planning commission in Thousand Oaks, in Southern California. Then he and his wife took a vacation trip, saw Clear Lake, and found it absolutely mind-boggling. He bought the Redbud Lodge in Nice, at the upper end of the lake, and joined the chamber of commerce. He had no intention of getting into politics, but then there was a question about his deed.

> That's when I saw my supervisor.... His response was that it was a state law; we couldn't do anything about it.
>
> My response was, "You represent us. Did the board of supervisors complain about it to the state, or at least write them a letter?"
>
> His response was, "Mr. Wilcox, the last time I wrote a letter to the state Assembly was seven years ago, and I'm still waiting for an answer."
>
> ...So I came home and told June [my wife] what happened, and she said, "Well, look, either do something to change it or accept it."...
>
> So I took her advice, and...I shaved my beard...and went down to the courthouse and filed...to run for supervisor of Lake County. (Wilcox interview, KMD vol. 7, 363)

He was elected in November 1978, and was seated in January 1979. He helped to bring the county's general plan up to date, to restructure the board of supervisors, and set up a viable planning department. He recalls learning of the mine project:

> We heard...they were nosing around Lake County, Napa, and Yolo, but it was up in the hills there.... So, really, the gold mine came softly.... They didn't come in like gangbusters.... We've always had controversy with Yolo Water District...and historically a tremendous resentment...that they stole our water. Well, they didn't steal our water. Water rights were open, and they took it. They applied for it.... They want nothing to touch "their" water....
>
> Our planning would work with Napa planning.... It was the bureaucrat working with the bureaucrat, and then coming before our respective commissions or boards and getting approval or disapproval. And if you got a disapproval from the commission, then you'd appeal to the board.... We're the last okay or yes or no.... I don't remember any big controversy at that level between the three counties [as relates to Homestake]. (Wilcox interview, KMD vol. 7, 371, 390, 391)

Ray Krauss recalled,

> In fact, I think it was Supervisor Walter Wilcox who said, "If you hire people locally, I've got no choice but to support your project because they'll vote me out of office if I don't." He was kind of a homespun, but very bright and sophisticated politician, and that was very, very true. To the extent that we hired local people who were voters in the community, then they were bound to support our projects. It wasn't rocket science. (Krauss interview, KMD vol. 8, 309)

THEY HAD A LOT OF PROPS

John Ceteras was born in Ohio in 1944, moved to California in the early 1970s, and in 1974 arrived in the Capay Valley and became active in the organic farm movement. He was a volunteer fireman, chairman of the Yolo County general plan committee, and the president and superintendent of the Rumsey Water Users Association, which held riparian water rights there dating back a century. He became a spokesman for those who feared what a mine would do to the environment they cared so deeply about. He recalled the meeting at the Guinda Grange Hall:

> [T]hey had a lot of props...they had their video cameras set up. They had a regular professional videotape company doing this for them. And they had all their executives: the mine manager, a couple of the vice presidents.... There were a lot of principals there from Homestake, and...they had their lawyers with them.... There were us in our jeans and overalls, and there were them in their Armani suits and Gucci loafers and their piles of documents, so...it really made people uncomfortable.... [T]here was a lot of discomfort by all this hoopla. (Ceteras interview, KMD vol. 2, 80, 82, n.p.)

THEY WERE JUST FEARFUL WHEN THE WORD 'MINE' WAS USED

Twyla Thompson, Yolo County supervisor from 1975 to 1985, was the granddaughter of a pioneer settler and rancher, and the family had continued to be ranchers on that land. She worked on the development of the Yolo County land use plan and was well acquainted with earlier controversial issues surrounding Cache Creek gravel mining. She knew what her constituents most cared about. There was great variety:

> I had all of the university [University of California, Davis] campus and student housing in my district...so it was an interesting group

> of diverse interests, from farming people to conservative people related to agriculture...retired people, people that had been in other professions that took up farming, people that had jobs and would commute—airline pilots.... A whole diversity of folks...that were very interested in the community and in the valley itself.... They were concerned over air quality, and...what would happen to the land.... [T]hey were just fearful when the word "mine" was used. (Thompson interview, KMD vol. 7, 122, 113, 133)

Twyla Thompson was appointed in January 1982 to serve as Yolo County representative on the Environmental Data Advisory Committee with the charge to "review and provide timely comments on the study plan and the baseline report" (Thompson interview, KMD vol. 7, 136). They met in Napa on February 10, 1982, after preliminary discussions with the planning staff. She was well-informed on the plans and commitments that Homestake had made.

In Yolo County a major concern was changing the ordinance to allow private reservoirs and retention basins and water transmission facilities to be used for mining activities, because until now they were only for farm ponds or cattle. Homestake wanted to build a dam on Davis Creek, to supply water during construction and later for processing. There were no residents in that area, and it was nearly inaccessible, as Twyla Thompson knew:

> You could get to the project through a back road over what we call Low Water Bridge...there was a back way that you could get from up above Rumsey over to Lower Lake...it's really a four-wheel-drive road.... And there is a road that would go off from there that you could eventually go back to the Homestake project. But those roads are not well-traveled roads, and only people with four-wheel drives would undertake them. (Thompson interview, KMD vol. 7, 127)

When the time came for the county supervisors to vote, Twyla Thompson's approval guaranteed that the application for 327 permits could be allowed.

> I've always made decisions based on facts.... [M]y role was to find out exactly what was going to occur, how it was going to occur, and what some of the safeguards would be in the project today, as opposed to the lack of safeguards in old mining operations in the past.... And I felt satisfied that this company really wanted to do it right. (Thompson interview, KMD vol. 7, n.p.)

KEPT THE FIRE UNDER THEM THE WHOLE TIME

One of the most articulate opponents of the Davis Creek dam in Yolo County was Will Baker, professor of English at the University of California, Davis (UC Davis). He was born in Council, Idaho, studied at the College of Idaho, the University of Washington, the Sorbonne in Paris, and the University of Hawaii, and received his doctoral degree from UC Berkeley in 1964. He taught English at Reed College in Oregon before going to Davis in 1969. He bought an almond farm in Capay Valley and was active in community organizations. He found that the old-timers welcomed changes more than the newcomers:

> [Y]ou had people who had recently...staked on this place their hope for their future and their dreams of what kind of life they were going to lead.... They were immediately more radical than most of the old-timers in terms of preserving things as they were, which is an irony pointed out to us often and somewhat acerbically by the old-timers, that we were more conservative than they were.... [They] believed it was good, straightforward American business enterprise at work and that it was going to benefit the economy. (Baker interview, KMD vol. 1, 69)

Yolo County's concern for water dates back a century, and Baker knew the history. By law, the top nine feet of water in Clear Lake belongs to Yolo County.

> They bought it from a bankrupt water and land company, who bought it from another water and land bank company who acquired it from a railroad and land company in the great nineteenth-century give-away. It's the story of the West.... [P]eople are so water-conscious there that anybody who impounds some and keeps water that would otherwise run off is going to be positively assessed. (Baker interview, KMD vol. 1, 81)

Baker went to the informational meeting in the Grange Hall in Guinda.

> [I]n the West, in California, if you raise the prospect of gold, for many people that triggers a pleasant blend of exhilaration and inquisitiveness and adventurousness.... [T]here were eyes shining in the room, I could tell, and the whole thing set off my alarm system.... [T]here was a little speech about how the days of corporate rapacity were long gone, and they were going to do this right...be good neighbors, and...do a clean project.

> ...[O]ne of the things I would tell myself when I would be sitting through these committee hearings or supervisors meetings or planning commission meetings, and here was this whole row of high-priced lawyers and experts in geology and engineering and so forth in their suits flown in from God knows where and their own slide projector and show...[and] I would think, Of course not.... Of course it's going to go, and eventually they will get their mine.... They were applying a lot of grease, and they did it well.... Krauss is a masterful PR [public relations] person. They ran a very good campaign.
>
> ...In terms of their trade and that context, I think it is a clean project and a well-run project and probably one with minimized impacts.... But I also think that a significant measure of that cleanliness and commitment to monitoring and restoration are the consequence of our having kept the fire under them the whole time.... Someday some historian is going to...say, Well, there was some opposition. (Baker interview, KMD vol. 1, 83, 88, 92–93)

SOME OF US REALLY ENJOY WEARING JEWELRY

The powerful environmental organization, the Sierra Club, did not endorse the Manhattan project, but neither did they oppose it. Ray Krauss recalled that Ada Mehrhof, one of the women who had toured the project with him, spoke for the Yolo-Solano chapter at the Yolo County planning commission hearing:

> [W]e have evaluated the environmental matters relating to this project, and we have decided that the engineering and the environmental engineering is sound. We have taken all of the documents and had them reviewed by people whom we believe are qualified to review them and whom we trust, professors on campus, and nobody has been able to point to any inadequacy in the engineering and design and environmental assumptions that have gone into the project.
>
> We considered the role of mining, in particular gold, in our lives and decided that we personally and as a society do need gold, that we use it in our...computers and have it in our teeth.... And some of us really enjoy wearing jewelry.... Given that we use it personally, if we were to oppose this project we would be in effect endorsing the production of gold elsewhere under circumstances that would be less environmentally and socially favorable. And therefore the conclu-

> sion was, we choose not to oppose this project." (Krauss interview, KMD vol. 8, 349)

YOU HAVE TO CONSIDER WHAT'S IN THE PUBLIC GOOD

In Yolo County, as part of the permitting decision, a technical review panel was established as a watchdog agency for continued environmental monitoring at the mine. Avery Tindell was one of the first members.[1] His grandfather, Joshua Avery, was city treasurer of Lead, South Dakota, home of the original Homestake Mine, in the 1890s. Avery Tindell was born in Oakland, California, and graduated from UC Berkeley in 1941. During World War II he served as a fighter pilot, and was awarded the Air Medal with Seven Stars, the Distinguished Flying Cross, and the French Croix de Guerre. He retired as chairman of the board of Pacific Marine Insurance Agency, and in 1980 bought a farm in Rumsey, in the Capay Valley area of Yolo County. It had a new well three hundred feet deep, five acres of oranges, and about twenty-six acres of almonds. The mandated general plan was being developed, and Tindell recalled,

> So I just wanted to be doing something constructive in my retirement rather than just sit in a chair and atrophy.... If you're part of the community, you should participate. (Tindell interview, KMD vol. 7, 184, 188)

Tindell says he was one of the few Republicans in the area. He soon served as subchairman of the public health and safety section of the general plan committee; John Ceteras was chairman.

From 1985 to 1992 Tindell served with Ceteras on the nine-member Technical Review Panel, appointed by the board of supervisors of Yolo County, to monitor Homestake's activities. They met monthly, reviewed the quarterly

1. Avery Tindell made significant donations to the Bancroft Library archives: large sections of the D'Appolonia environmental report; the Bay Area Air Quality Management District report of July 9, 1984; "Summary of Analysis, Homestake McLaughlin Project, Permit Application 29351"; Napa County Conservation Development and Planning Department annual report/review of the 1987 environmental monitoring program, McLaughlin Project, May 1988; Rumsey General Plan, 1975; Environmental Data Advisory Committee report on review of the environmental monitoring plan, McLaughlin Project, November 1983; Yolo County Planning and Community Development Agency report on review of baseline data adequacy, environmental quality standards, and enforcement—McLaughlin Project, February 1984; as well as miscellaneous correspondence concerning the Yolo County Technical Review Panel of which he was a member.

reports submitted by Homestake, and compiled their own annual report. He and Ceteras collaborated again.

> John and I both had a great interest in it, attended practically all the meetings.... I had had organic chemistry, inorganic chemistry, and...other stuff.... [M]y position as a Republican was, as long as they complied with all the laws and the environmental concerns, I didn't have any objections.... You have to consider what's in the public good.... And it was very well done. (Tindell interview, KMD vol. 7, 193, 199, 205)

John Ceteras recalled the technical review panel:

> Well, we fought for it, and...through a lot of going to the meetings and lobbying everybody, we won, got the...concept approved... [and] also, that we would have a public member. But we had the air quality,...the animal life,...the invertebrates and all the different disciplines, the hydrogeology, fish life, aquatic biology, all these bases covered by someone from the university primarily.... And then the public members...myself and Avery Tindell, were appointed by the board...the condition...that they put...That we would be able to monitor the results of their monitoring, on an annual basis, and report to the planning commission.... Initially, it was for twenty-five years, which was the expected life span of the mine.... [T]hey promised the moon and the stars...and we wanted to not just take their word for that.... "It's no question in my mind that it has kept everybody honest."...I think we did good...as far as being able to be assured in the long run that we weren't going to be harmed by this mine. (Ceteras interview, KMD vol. 2, 93–110)

Ray Krauss persisted through more than a year of discussions, forecasts, analyses, commitments, plans examined and changed, public information sessions, on-site tours. Finally, on August 27, 1982, he filed applications for 327 permits; more than a year later, the permits were issued.

In early January 1984, the *Napa Valley Register* carried a front-page article by Doug Ernst, staff writer, with the headline:

> **Supervisors Uphold Homestake Approval Napa County**
> Supervisors today unanimously upheld last month's vote by county planning commissioners to issue use and mining permits for Homestake Mining Company's McLaughlin Project near Knoxville. Two Yolo County residents appealed the December 2 commission decision, alleging that Napa County did not allow adequate review or comment and that environmental information was incomplete. In rejecting the appeal, the board ended Yolo residents' hopes of stalling the review process in Napa County. (Ernst 1984)

Finally, the Manhattan Mine project became a reality. Each county would have a legacy: in Napa County, the pit would become a lake; in Lake County, an improved Morgan Valley Road would pass a waste disposal area turned into a wetlands and wildlife habitat; in Yolo County, a reservoir would store clean water for irrigation and firefighting. When mining operations ended, the University of California would have a new research station for the study of serpentine ecology.

Jack Thompson recalled, "This is how long it took to get the permits and in retrospect, you say, that was good. It could have been a hell of a lot worse" (Thompson interview, KMD vol. 7, 42).

William G. Langston was Homestake vice president and general counsel for nineteen years, from 1973 to 1992. He believed that reconciliation of environmental concerns and industrial activity is one of the great issues of our time, and that history will show that Homestake has fulfilled every commitment made in that important cause.

PART 4

CHAPTER 15

THE CONSTRUCTION PHASE

We engineered the dickens out of everything up there.

—Rex Guinivere, construction manager

When he assembled the new team in 1981, Bill Humphrey knew that as soon as the permits were granted, he would need an experienced construction engineer. Rex Guinivere was hired on September 1, 1981, to manage the design and construction of the project. He was born in Auckland, New Zealand, attended the University at Auckland, and in 1956 he graduated from the Otago School of Mining and Metallurgy, the oldest professional school in New Zealand. Then he spent, as he says, "six years in the bush in Australia, and three years in the jungle in Malaya," where he was production superintendent for Eastern Mining and Metals Company (EMMCO) at Bukit Besi, an 80,000-ton-a-day open-cut iron ore mine. He came to the United States on holiday to visit a brother who had worked for Kaiser Engineers in Australia; after obtaining the necessary papers, Rex stayed on to work for Kaiser on the Hellhole Dam on the American River, in California. He later transferred from construction into the newly created minerals division, and by 1981 was Kaiser's vice president for nonferrous minerals (Guinivere interview, KMD vol. 3, intro., 246).

Guinivere spent the first months at Homestake just reading files. Like Bill Humphrey, he recognized the situation:

> I had been brought in to manage a project, and I was...allowed to attend their meetings, but...My people were sitting there at the command of the geologists.... [E]verybody was sort of kept well abreast of what was happening.
>
> But you get into this situation, you get operations people who don't think they need engineering people, and you get exploration

> people who don't think they need the operations people or the engineering people.... It's really a bureaucratic mess.
>
> ...We had no intention of building up a little hierarchy. There was never more than six or eight people in the engineering group. (Guinivere interview, KMD vol. 3, 259)

In May 1982 Rex Guinivere was finally put in charge of construction for the Manhattan project; it would be groundbreaking in every sense of the word. The team included John Ransone as project engineer, Bob Lear as manager of the metallurgical group, and Bob Previdi as manager of the mining evaluation group. A consultant had predicted employment of 440 construction workers; in the end, there were more than 1,200 at the peak period. From the start, Guinivere wanted to do it as a nonunion project, and officials at Homestake, traditionally a nonunion employer, agreed. He recalled,

> [The] construction people...said, "You've got to be nuts.... You're going to try to do a big non-union construction project seventy miles out of San Francisco, right next to Konocti, where the Plumbers Union has got their big fancy resort?"
>
> I said, "Well, with the economic conditions we have at present, if I can't do it now, I can never do it."... [B]ut it wasn't easy. It took a lot of work and a lot of training.
>
> ...This was the biggest nonunion construction project that had ever been done in California. We did 3 million man-hours.... It was thirty-odd miles long.... [The contractors] were all nonunion...with one exception.... [W]hen we came to lift the cold box on the oxygen plant, we had to bring in Bigge Construction, a union company, because they were the only people who had the cranes big enough to handle lifting a 125-foot cold box up without breaking it.... It was only the top engineering and maintenance people who were brought in from outside...60 percent of our 3 million man-hours was local people. (Guinivere interview, KMD vol. 3, 273, 274, 275, 276, 278)

The company made a commitment to Lake County that over half of the workers would be hired locally. Some construction companies doubted that there were enough capable local workers and declined to bid on the project for that reason. The McKee Corporation of Cleveland, Ohio, originally a chemical engineering company, had merged in November 1978 with Davy

International, Ltd., of London, England, to form Davy-McKee Corporation.[1] They were awarded the engineering contract on June 1, 1982, and opened an office in San Ramon, California. Construction at the mine site began on a limited basis in September 1983. On January 20, 1984, the first concrete was poured.

Because so many of the local hires needed training, Yuba College, the local community college, expanded to include more classes in work-related skills. Davy-McKee donated equipment and people, and Homestake donated materials to build a shop for training classes for welders, electricians, helpers, carpenters, and concrete form work. The first location was in a hangar at Pierce Field, and Homestake later helped to pay the college for an industrial arts center in a permanent building in Clearlake. The company worked with the Employment Development Department in Lake County to recruit people for training. They gave preference to people who were able-bodied and unemployed, and anybody who completed the course was guaranteed employment during the construction phase, and later in the operations. Guinivere recalled,

> There was something like 30 percent unemployment, and it was essentially a welfare community and an old folks' home, mostly people living in trailer homes and whatnot. We were giving people their first full-time job in their life, and they were in their thirties.... [T]hey're still getting excellent pay. But they're not paying union dues, which are quite considerable. (Guinivere interview, KMD vol. 3, 275, 277)

Homestake also committed to compensate the local schools for the extra load placed on them by the influx of workers with school-age children. They hired an expert to estimate the impact, and Thompson recalled that the impacts were less than predicted for the construction period, but more than predicted for the operating period.

1. *Chemical & Engineering News*, October 02, 1978. A longtime, presently ailing major U.S. company in chemical engineering and construction, McKee Corp. of Cleveland (formerly Arthur G. McKee & Co.), soon will become part of a British engineering and construction firm, Davy International Ltd. of London. McKee's board of directors has approved a proposal from Davy International to make a tender offer to buy McKee's 3.2 million shares for $33 per share for a total purchase price of $106 million. Success of the tender offer would result in McKee's merging into a subsidiary of Davy International.

John Drummond, attorney for the Lake County Schools, negotiated the mitigation agreement:

> [T]hey agreed to pay a sum of money for every child that was generated to become a pupil in the school.... [T]here would be an annual survey, and then the students that were identified in the survey as having come to Lake County as a result of Homestake's activities were then considered to be the basis of payments to the school districts. And over the years, Homestake Mining Company made payments to the school districts in the amount of approximately $741,000 total over the years, based upon the number of students that enrolled in the school district as a result of the mine going in. (Drummond interview, KMD vol. 2, 272)

Bonny Hanchett, editor of the *Clear Lake Observer*, assessed the local impact:

> So it had a very positive impact.... There was no lodging at the mine. So they found places to live, and that helped the resorts; and, of course, they had their payroll, and that was quite a substantial payroll because all of the processing went on in Lake County, so most of the people that worked in that end lived in Clearlake or Lower Lake. (Hanchett interview, KMD vol. 4, 86)

Homestake paid the town of Lower Lake $160,000 to compensate for upgrading the streets and utilities to cope with the extra traffic. A traffic light was installed at the intersection of Highway 29 and 53, and a left turn lane. A parking lot was put in on Highway 53, where construction workers could park; they were taken to the worksite on old school buses that were leased from a private corporation in Lake County.

Claire and Herbert Fuller, both Navy veterans, moved to Lake County in 1982. He was a butcher; they already owned and operated small grocery stores in Clearlake and Oroville and bought the Economy Market on Main Street in Lower Lake, renaming it Fuller's Superette. Main Street runs east from the intersection of Highway 29 and 53. In those days, it went past a thrift shop, a barber shop, an old wooden theater, the Five Brothers saloon, a mortuary, and then the Superette, at the edge of town, where Main Street becomes Morgan Valley Road, State Route 140. Homestake, eventually, had an address at 26775 Morgan Valley Road.

Claire Fuller recalled an immediate effect of the mine on her business:

> And one of the...women, that worked at the store when we bought it owned property out there, about seven miles from town, and Homestake bought it from them.... Out there they didn't have any electricity.... In fact, he used to hook up the TV some way to the battery of his car so he could watch television.... [S]he was in her sixties then, and her husband was getting ready to retire, and she wanted to move into town.... As soon as Homestake paid them for the land, she quit working. (Fuller interview, KMD vol. 3, 130, 131)

Then, when construction began,

> The guys would come in to the store after work, buy beer, and some of them cashed their checks there at the store, and we'd seen them going back and forth.... They built a parking lot for them...next to [Highway] 53,...and then the guys would park their cars there, and they would pick them up in buses and take them out to the mine. So the buses were back and forth, three, four times a day.
>
> ...I'd say probably from '81 to '84 were good years.... The men would stop on their way home...we used to make sandwiches [for their lunch]. Five Brothers [bar] was pretty good then.... You would go uptown in those years, and you couldn't find a place to park. (Fuller interview, KMD vol. 3, 129, 132)

Another local merchant, Beverly Magoon, recalled the impact in Lower Lake:

> When they started doing construction...[it] was just a really busy time.... That was when all the buses were going back and forth, so it was pretty busy for quite a while there. It employed a lot of local people.... Some people had more money than they had before because they were working in construction there.... They bought some things from us, but...I think it was more for the bar...than it was for us.
>
> ...[T]he general feeling of the town, I think, was really good then.... [T]he mine put in a road. That made a lot of difference. But...the road was torn up for a long time, so that wasn't very advantageous...so there's kind of that sort of positive and negative things at that time. (Magoon interview, KMD vol. 5, 196, 214, 215)

Ray Krauss recalled positive effects in the community:

> But it certainly changed the character of some of the local communities. I know there were a number of little businesses that popped up that served the construction crews in terms of coffee in the morning and box lunches, and the Five Brothers bar at night. They had a huge check-cashing enterprise in the back room, to cash people's paychecks. So the local economy very quickly adapted to and took advantage of all the opportunities that were presented.
>
> ...[R]eal positive demonstration of community acceptance, was...the Christmas of '82.... The Lake County Economic Development Board, which included Dave Hughes and Pam Lubich, two or three people prominent in the Lake County community, who had been appointed by the board of supervisors to be the community development board or the county economic development board.
>
> And they gathered together Christmas baskets with local Lake County products. They had pears and walnuts and wine...and put Christmas ribbons on them, and they made enough for everybody in the San Francisco office, and they drove them down to San Francisco, and they had this gathering of all the San Francisco personnel in the board room, and they handed out a Christmas basket to everybody and said, "We really want to welcome you to Lake County. We're really excited about your joining our community." (Krauss interview, KMD vol. 8, 327, 357)

The company spent $23.5 million on exploration and preliminary environmental and engineering work. This included the table-top process model built by Davy-McKee that cost around $300,000, an expense that was justified because it would avoid costly surprises later. Some of the work for this model could be done by computer, but much of it needed painstaking drafting by hand. The miniature components, such as tanks and pipes, were purchased and shaped according to the plans. When completed, the model showed the processing equipment, all color-coded: green for electrical, other colors for the various circuits. The engineers knew exactly what the plant would look like when it was completed, and the model guided them through the construction. When the operating phase began, it was a valuable training tool for new employees.

John Turney, metallurgist, recalled the training:

> [E]very valve and every pipeline is on that model—so you talk about, "Okay, how are we going to fill this tank, and what are we going to do?" You walk up to the model and say, "Well, open that valve there," and point to it on the model, trace the line along and say, "Then you'll discharge into this tank here, and when that tank's full it'll overflow to this one." You would spend your time looking at the model, and then you would actually go out into the plant—it was under construction then—and you would look and see where everything was and where that valve was.
>
> ... [T]hey interviewed people from Lake County who had perhaps worked out at the property during construction. We'd say, "I think this person might be able to learn and be a good operator." That group was brought in as well, and they had never operated anything like this before, so you had to run through some very fundamental issues like how to turn a pump on and off, what priming the pump meant, what to listen to in the pump, the safety considerations of dealing with high voltage, and what switch gear meant, what motor control centers are, what a control room was, what a control valve was and how it worked, how did a controller know to fill up a tank and then stop filling up—how did that actually work? So the fundamentals were there, people understood the simple things in their homes—how their toilet bowl works—so you were able to use that as a starting point to explain how you're using the same techniques in an industrial environment to control a process. (Turney interview, KMD vol. 7, 285, 286)

Guinivere was distressed by some of the design work that had been done by a Canadian company:

> But I had been in engineering design and construction work for a long time by this time, and... [they were] giving stuff to me that I just couldn't accept.... [W]e paid them a quarter of a million dollars for a pile of paper that was almost worthless. (Guinivere interview, KMD vol. 3, 265)

He learned about the plan to use autoclaves in the processing:

> No one had ever done whole ore autoclaving.... I initially was uncomfortable with it until I looked at all the data and realized that

> this was the only possible way to go.... [Y]ou were faced with...a large deposit of about 3.5 million ounces, with gold that was mostly refractory.... [T]hese small particles of gold—and it was submicron particles—they were surrounded by refractory sulfides of silver.... What you had to really look at was reversing the thermodynamics of the hot springs that had placed them there. (Guinivere interview, KMD vol. 3, 265)

He was not the only skeptic:

> Just to let you understand what the attitude in the industry was to what we were doing:...It must have been—oh, I guess it was 1985... we were building the autoclave system, and we weren't expecting to start that up until August...Plato [Malozemoff] was chairman of Newmont, and...Plato was the speaker at the mining dinner at Berkeley, at the mining school at the university there. I was there, and Bill Humphrey was there, and we were standing in this little circle before the dinner, having a drink. Plato...leaned over to Bill and said something to him and then walked away. Bill sort of looked glum, and shrugged his shoulders, and I said, "What did he say?" He [Bill] said, "[He said], 'You'll never make it work.'"...
>
> I just laughed, and I said, "Well, he's going to be surprised." (Guinivere interview, KMD vol. 3, 266, 267)

Guinivere was frustrated by the regulatory requirements:

> It was just one terribly complex project from beginning to end... because we were...involved with three counties,...the Bay Area Air Quality District, the forestry [U.S. Forest Service],...the BLM [Bureau of Land Management], and the EPA [Environmental Protection Agency].... It was just unbelievable red tape. (Guinivere interview, KMD vol. 3, 261, 270, 292)

For example, at the mine site, dust control was an issue, and several air quality monitors were installed, each one costing $150,000:

> It was so bad that during construction, I estimated at one time we had something like 45,000 horsepower of big construction trucks and dozers and scrapers working on the excavation work, and the preliminary mining at the mine site. We were using the rock to make the gravel for our concrete for construction purposes, and we had

> a gravel plant in the mine site. That had an 850 kilowatt generator, a diesel generator. That was a stationary plant, and it came under the aegis of the Bay Area Air Quality District, and we could not operate that...as much as we wanted, despite the fact that it was surrounded by, as I say, about 45,000 horsepower of the same sort of diesel equipment, spewing out these diesel fumes, but that stationary plant came under the aegis of the Bay Area Air Quality District, and they wouldn't let us operate it more than so many hours a day. (Guinivere interview, KMD vol. 3, 270)

Ray Krauss also recalled the difficulties of permit compliance:

> And keeping a construction force with two major contractors and dozens of subcontractors and huge numbers, pieces of earthmoving equipment—all in compliance with all of the conditions of the permit all of the time was a real challenge.
>
> ...[T]hose were obviously things that the construction managers found vexing, but issues that were required by law to be...properly addressed.... But it was almost a cops-and-robbers relationship. I was never comfortable that if I weren't there, looking over their shoulders, that they would...comply with what they knew to be the requirements of those permits; that if they could get away with it, they would get away with it.... I had some real battles with them over the construction period....
>
> Fortunately...Jack Thompson and Bill Humphrey and the other folks at Homestake were really committed to assuring that our environmental record was sound.... But it was a busy eighteen months to make sure that we stayed on track. (Krauss interview, KMD vol. 8, 329, 331, 341)

Jack Thompson recalled benefits that the construction brought to the community:

> We needed a lot of water for construction for keeping the dust down, compacting things, concrete, all those kinds of things. We laid a pipeline alongside the road all the way from Lower Lake and bought the water from them. Then the portion that was inside Lower Lake, we did to full public utility standards.... They had a fairly undersized system with a lot of problems, and we improved their intake and put in a new water main through town.... The old one was leaking.

> ... [A]nother one of those local benefits derived from the project ... [w]e built a 115-KV line, which was the big line for our plant. Underneath it we put cross T's and put up a smaller line for local distribution, so the people along the road got power for the first time. (Thompson interview, KMD vol. 7, 91, 92)

For a long-term source of water for the process plant, a small dam was planned on Davis Creek, and this came under the jurisdiction of still another agency, the California Division of Safety of Dams. It became a major challenge for Guinivere:

> Most places where you get a river or a creek in California, it's going down a fault. There was a fault zone on the left bank.... They [Division of Safety of Dams; DOSD] decided that we should do a lot of grouting on the dam. They also decided that we had to cut out all the fault zone on the left bank and put in concrete pedestals all the way up.... The grouting cost us $2.5 million, instead of the $55,000 that we had budgeted. (Guinivere interview, KMD vol. 3, 278, 279)
>
> Weather was another challenge:
>
> I've been around a lot of dams, and it's axiomatic that when you are building the dam, you will have the thousand-year flood, and when you're trying to fill the dam, you'll have the thousand-year drought. (Guinivere interview, KMD vol. 3, 281)

Ray Krauss recalled a story he was told by John Ransone, Homestake engineer:

> When the dam was partially constructed and we had this nineteen inches of rain in October, we actually had water accumulating behind the partially completed dam. One afternoon we had two inspectors on site. One was from the Division of Safety of Dams, and the other was from the Department of Fish and Game.
>
> The inspector from the Division of Safety of Dams said, "You must release that water. You're not permitted to impound any water behind an only partially completed dam."
>
> ...And the man from the Department of Fish and Game said, "You must not release that water because it has a high sediment load in it and if you release it you'll violate downstream water quality standards and damage the riparian resource."

And John said...he looked from one to the other, and finally he said, "Well, the warden from Fish and Game is carrying a gun, so I'm going to do what he tells me to do." (Krauss interview, KMD vol. 8, 333)

Guinivere argued that they did not need to build a pilot plant for the autoclaves:

And I said to the board and to Harry [Conger]...that rather than spend $7 million on a pilot plant, we thought that we could spend some extra money in engineering.... We had a mini-pilot plant, one or two kilograms a day...up at Fort Saskatchewan. And they did such good test work, and then John [Turney] and his people and Davy-McKee and the Vaal Reefs people did such good engineering work, that that project started up and did exactly what it was designed to do.... Everybody in the industry was waiting for us to fall flat on our ass.... [And] now everybody does it. (Guinivere interview, KMD vol. 3, 285, 291)

Jack Thompson recalled a brief attempt by unions in 1984:

The only push was during construction and, interestingly enough, they couldn't generate much local support because of our emphasis on local hiring and that we had delivered the jobs.... So they imported bus loads of picketers, and they picketed our mine, and most of the workers crossed the picket line. Only a small handful didn't. There was a lot of communication back and forth, and when these guys from out of town found out a few of the facts of what was going on, they lost a lot of their picketers, who left right away. Then there was just a small core group of picketers that was probably less than ten or twelve. It lasted a week or so, and then they were gone. Very ineffective effort. (Thompson interview, KMD vol. 7, 106)

He praised the safety record:

[O]ur accident frequency was about half the national average, and the severity was about half the national average, so it was a safer than normal construction project.

And then the operations went a whole year before we had a lost-time accident, which was quite an achievement. I attribute it primarily to the training—all that extra heavyduty training that we did. (Thompson interview, KMD vol. 7, 108)

The start-up at the processing plant in 1985 went well, as Guinivere recalled:

> And we expected to start that up in August...run it for a couple of weeks, shut it down, and then tear it apart and take a look at it and see what had happened inside.... No one knew what was going to happen with the materials in [the] construction [phase] because no one had done what we were doing with a heavy slurry and high pressure and high acidity.... We spent August tearing it down...and we couldn't find much wrong at all. We started up...operated it, it worked exactly as planned...and in September, we handed it over to the operators and said, "Here, it's all yours," and walked away from it. (Guinivere interview, KMD vol. 3, 302)

He summed up his feelings:

> It was the best thing that I had ever been associated with.... I think it was [satisfying] for everybody that was on the project.... We did things we never believed we would ever be asked to do, but we had to do.... We engineered the dickens out of everything up there. (Guinivere interview, KMD vol. 3, 307, 312, 316)

CHAPTER 16

AUTOCLAVES: A GLOBAL EFFORT

I mean, it was something, when they brought the autoclaves right down Main Street.

—Claire Fuller, owner of Superette Market

That's what the real story of McLaughlin is about. Just because we've got some ground and it's got some gold in it, it doesn't mean we have a mine.

—John Turney, Homestake metallurgist

In January 1985 Claire Fuller in Lower Lake, California, saw the culmination of years of investigation around the world. An important legacy of the McLaughlin Mine project was the continuous-pressure oxidation system that beneficiated gold ore without harm to the environment. It has been widely copied and resulted from worldwide research and procurement by the Homestake team, several of whom came from other countries.

John Turney, a metallurgist who worked on development of the acclaimed process to treat the McLaughlin Mine ore, was born in Queensland, Australia. His grandfather and great-grandfather had been associated with Australian gold mining; his father was a graduate of the Bendigo School of Mines. John graduated from Monash University in Victoria with a degree in chemical engineering, and in 1977 earned a master of science degree in metallurgical engineering from the Colorado School of Mines, in Golden, Colorado, with a thesis on high ionic strength solutions, especially extraction of lithium from brines. He worked first for AMAX in the environmental research group in Golden.

In January 1981 he began work with Homestake's metallurgical research group in Golden, Colorado. The McLaughlin ore body had already been delineated and preliminary drilling had been done. Richard Kunter was also on the team; he was an American who had worked with pressure and autoclave systems at Western Mining and Kanowna in Western Australia.

Robert Pendreigh, a consulting metallurgist formerly with the firm of A. H. Ross, of Toronto, Canada, was brought in. He had worked in South Africa, at the Witwatersrand mining district, where a continuous autoclave process had been used, with ore in a dilute solution.

Turney recalls,

> I'd say they were at the point where they said, "Well, we might have a problem here, because a lot of this stuff is refractory."...Traditionally, what everybody did, if you had a refractory sulfide, you'd put it in a roaster, you'd heat it up, and break down the sulfide with heat, which would then evolve sulfur dioxide, which goes up a stack. (Turney interview, KMD vol. 7, 211)

Concern for air quality, however, now prohibited this. An option was to treat the sulfides under pressure with oxygen, which had been done successfully in a laboratory.

> I think everybody was quite acquainted with the fact that you could oxidize sulfides and essentially increase the oxidation rate of sulfides if you did it under pressure in a pressure cooker. But the question was always, how could you do it?
>
> ...I think people knew you could do it...but whether that was practical, everybody would have said no... [b]ecause the equipment wasn't available to do it. That's really what part of the real story of McLaughlin is about. (Turney interview, KMD vol. 7, 233, 234)

A formal research program was set up at Hazen Laboratories in Golden, Colorado. Turney led the investigative team that included Homestake employees John Ransone, Andy Sass, Klaus Thiel, and Roger Madsen; they traveled to Europe and South Africa in search of the right ideas, materials, and equipment. Turney says, "The key to the operation and perhaps the essential component of the innovation was the development of the let-down system. Roger Madsen, the mechanical engineer on the project, deserves credit for this achievement" (Turney interview, KMD vol. 7, 280–282).

Roger Madsen was born in Hibbing, Minnesota, in 1923. His father was the master mechanic for the International Harvester Company at an openpit iron mine, and Roger went to work in the mine immediately after graduating from high school. He was on the track gang, the rail track in one of the huge open-pit iron mines of that region. He worked and also attended Hibbing Junior College until he was drafted; he served for three years in the Army Corps of Engineers in Europe. After he graduated as a mechanical engineer from the University of Minnesota, he began work at the Homestake Mine in 1948. He transferred to the uranium operation in Grants, New Mexico, and achieved recognition there for designing a successful skip loader. The Grants operation also used autoclaves, and the valves were a challenge there, too.

Madsen recalled his experience at the uranium mills in Grants, New Mexico:

> The treatment of the ore consisted of grinding the ore in ball mills and mixing it with a chemical slurry and getting it to the right consistency and putting it into autoclaves where air and heat would increase the rate at which the uranium would go into solution.... When you have the slurry coming out of the autoclave, or a series of autoclaves, someplace there has to be a pipeline and usually a valve of some kind where the slurry goes through the valve into a zone of much lower pressure. (Madsen interview, KMD vol. 5, 172, 173)

Because of his expertise, Madsen was called in to help with the design of the autoclaves, and particularly the let-down valves, at the McLaughlin Mine, where the slurry contained 40 percent solids. The investigative team visited a ceramics plant in England, and factories in Germany, Holland, and Austria that made slurry pumps and autoclaves. At one plant, they saw an autoclave with a let-down valve that had to be replaced every two hours. In Madrid, Spain, there was another that wasn't right for a continuous process. They looked at autoclaves at the Outokumpu stainless steel plant in Helsinki, Finland. Finally, they visited Vaal Reefs in South Africa, and here Madsen saw what was needed:

> If you change valves you have to shut the whole thing down. It takes a day to shut it down and two days to get it started up again. The construction of that valve and the wear properties of it at that time were overwhelming, at least until Vaal Reefs pointed out some of these things to us.
>
> ...They had designed their own valve. It was my understanding that they had solved their metallurgical problems in about three or

four months. It took them five years to solve the mechanical problem, this valve. They showed us the valve and they explained what they did, the design of it and everything. It was completely different from anything I had ever seen before. The material in it was what you would use in a grinding wheel, only better. It was not as hard as diamond, but nevertheless comes pretty close to it. They had developed a process for making the valve.... It was a molded silicon carbide...a material quite similar to what is used in grinding wheels, only more so.... They had marvelous machine shops and things like that over there where they could do these things.... [W]hen I asked them how long the valve lasted they said, "I don't know. We haven't worn one out."...From my viewpoint they had solved this problem. (Madsen interview, KMD vol. 5, 181, 194)

Madsen's common sense contributed to the decision to put the autoclaves under a roof:

This meeting was to decide what type of autoclave was going to be used. The decision at that time was the steel one that was brick-lined. These autoclaves at that time were designed for outside, to be outside, no building over the top. I sort of shuddered at this.... [T]he fellow from the German brick company...was giving a sales spiel about bricks and what you had to do, a lot of nice things about brick lining and things like that. When he got through I asked him... "What happens when it rains on this thing?"

...He almost leaped out of his skin. "Rain on it! You can't do that!"

...After the meeting we are out in the hall and Andy Sass comes over to me and says, "You know, you just added a million dollars to the cost of that autoclave right there." Well, if they didn't have what they needed it wouldn't have worked. They would have had nothing but trouble. (Madsen interview, KMD vol. 5, 183)

John Turney was involved in the crucial and unusual decision not to build a pilot plant first:

You'd be better off spending extra money in the full-scale plant, put some extra dollars into the plant, so that you could insure that it would work, rather than spend it on the pilot plant....

What we then ended up having to do was, the sizing of the letdown system and the autoclave, basically the thermodynamics, the

> heats of reaction and that, we did out of the back of the book. We basically pulled out a thermodynamics book and worked out how much heat would be generated, and the calculations that we did based on that theory proved to be fairly close to our numbers that we measured two or three years after start-up, where we actually physically measured temperature in and out, worked out heats of reaction, and came up with parameters.
>
> Those numbers now have become standard. The industry is using that as a standard number to design these things. We didn't have that. We had to work it out. (Turney interview, KMD vol. 7, 211)

Richard Kunter and John Turney received a process patent for the autoclave operation:

> What we patented was essentially the process, the linking together of pressure oxidation with gold recovery, and the conditions under which you could do that. So it's really the concept; that was what was patented. No physical hardware was ever patented. (Turney interview, KMD vol. 7, 211)

Turney helped to educate the public about the autoclave system:

> In fact, when I talked...to people, the general public,...I always use the analogy of a pressure cooker at home.... [T]o cook potatoes, you put them on the top of the stove, on the burner, and you would cook them to reasonable tenderness in thirty minutes. You put them inside a pressure cooker...they cook in ten minutes.... So I always used to say on the public tours, nature is oxidizing the sulfides in millions of years, and what we do inside the pressure vessel is the same thing, but we do it in eighty minutes. (Turney interview, KMD vol. 7, 251)

The final plant design was for three autoclaves, each one fourteen feet in diameter and fifty-two feet in length. The outer shell was 1.5-inch-thick steel, lined on the inside with lead and two layers of fire brick. Each was forged in one piece in Germany, then shipped to the Netherlands for lead lining. From there, the autoclaves were shipped to the port at Sacramento, where they were placed on barges to go to Napa.

The *Napa Valley Register* reported on November 3, 1984, that later in the month Homestake officials would request permission from the Napa County commission to use county roads to transport the first of three autoclaves to the processing plant in Lake County. They would follow the Silverado Trail, a well-traveled, paved highway on the east side of the Napa Valley. From there they would go on a steep, winding route along the flank of Mt. St. Helena, to the town of Lower Lake, and then on Morgan Valley Road to the mine. Homestake had previously strengthened and paved this last section to support heavy construction vehicles.

The same newspaper on January 31, 1985, ran a photo by Richard Mason and an article:

> Motorists along State Highway 20 in eastern Lake County were inconvenienced earlier this month as the first of three autoclaves was delivered to the Homestake Mining Co. McLaughlin gold mine. Pushed by one tractor and pulled by another, the autoclave was delivered by B&Y Heavy Haulers at speeds as low as one mile an hour on steep grades. Total weight of the rig was 400,000 pounds. (Mason 1985)

A later article noted that the trip from Sacramento took two weeks at an average speed of eight miles per hour, and that they were the heaviest legal loads ever transported on California highways.

Claire Fuller recalled that day when they went through Lower Lake:

> Oh, they're as big as this house.... They moved about five miles an hour, I think.... [A]ll the electric and telephone wires were back and forth across the street; they weren't underground like they are now, they had to take all those wires down so that they could bring these big autoclaves through town. I mean, it was something. Everybody just—we just stood and looked out the windows there at the store and watched them go by. (Fuller interview, KMD vol. 3, 114, 129)

Once the autoclaves were in place, ceramic workers came from the Netherlands and installed the fire-brick lining. It would still take months of trials before the autoclaves could begin operating.

CHAPTER 17

1985: START-UP

We call it de-bottlenecking.

—Patrick Purtell, maintenance engineer

On March 4, 1985, a select group including Homestake officials, Bill Wilder, and all the construction engineers met at the processing plant to watch the pouring of the first gold bar, and posed for photos. As Jack Thompson recalled,

> So we finally got it built. The first ore through the plant was in January of 1985. We started up the plant in two steps. The first was the non-autoclave portion of the plant.... We ran oxide ore through the grinding circuit, which was amenable to direct leaching. It just wasn't very much of it, but we had enough to start the plant with. So we...poured the first bar of gold on March 4, 1985....
>
> A lot of people don't know it: we actually did pour one two days earlier, but for practice because we melted it again! "Well, we'll make sure things work."...So then we continued the construction of the autoclaves. They were brought in, and everything was completed September of '85. By then, we had kind of worked the kinks out of the rest of the circuit, and the autoclaves started up. We could focus on that and get that going. (Thompson interview, KMD vol. 7, 89–90)

Patrick Purtell, maintenance engineer, recalls January 10, 1985, when the first ton of ore was dumped in the crusher. The start-up was hard to define:

> But there's a period before that where a lot of the process, the piping and some of the equipment is wet-run, wet-tested. They pump water through them to make sure there's no leaks in the pipes, that all the

pumps are turning in the right direction, that all the equipment will rotate like it's supposed to. All that is pretty well-defined before you put the product in, and that's part of the start-up process.... That was a cloudy area. [Davy] was actually in charge at that time.... All this pumping of water, no leaks in the pipes, everything turned in the right direction, when you pushed this button the right piece of equipment started. All that was done.... I'd say it was almost a year...[that] was considered start-up.

We had stuff that would start breaking. Had to change that equipment as you go. You keep going on. Keeping everything running...but at the same time taking out all of the little bottlenecks that were building up.... [We] call it..."debottlenecking." You just start going through one step at a time until you get the whole place running as smooth as it will go. And everything will supply the solutions you want, the ore you want, in the right places, and that's the whole gamut of start-up. So the end of it is real fuzzy. (Purtell 1999, 45, 46, 47)

John Turney recalled a problem at the start-up of the autoclaves:

And we started up one autoclave at a time. The first one that was started up was called C Autoclave.... There was a couple problems there. [T]he units are all brick-lined, so you had to soak them in acid inside to condition the bricks before you put it into operation.... [I]nside the autoclave all the wetted parts are made of titanium: the agitator shaft, the disk at the bottom of the agitator, and the blades.... [W]hen we emptied the unit, we noticed there had been degradation of the titanium metal...and the only way that could have happened is if there was some metal in there. Somebody made the suggestion that maybe a ladder had got left in there, and everybody said that was impossible, that we had checked everything. But the plastic remnants of the ladder were still in there, so we knew that there had been a ladder in there.... An aluminum ladder sitting in sulfuric acid generates hydrogen gas. Hydrogen is a reductant, so it actually reacted with the titanium surfaces inside there in the tank and dissolved some of the metal. The ladder, of course, totally disappeared. That was something that should never have happened. To this day we don't know why that ladder got left in there, but it was a problem we got by. (Turney interview, KMD vol. 7, 289, 290)

Davy-McKee's design was correct in most cases, but as Ray Krauss recalled, they hadn't anticipated

> the viscosity of the autoclaved ore...so that in order to move it through the screens it had to be diluted...[and that] meant more water, [and permits had to be amended] to allow us to recycle water from the tailings pond.... We added new pipes, new pipeline. (Krauss interview, KMD vol. 8, 354, 355)

Ray Krauss praised the start-up training:

> We did have an excellent start-up. The company brought on the operations employees quite a number of months before they were actually required to run the plant, and provided them with very comprehensive training. We had autoclave operation center and control center simulators and so they were able to simulate the operation of the autoclaves and...their response to upsets. It really was an excellent, smooth start-up. (Krauss interview, KMD vol. 8, 342)

Krauss and his team planned a community celebration at the mine on the weekend of Saturday, September 28, 1985, to cap the start-up. The official dedication was the first blast, and most of the Homestake Mining Company directors and officers were there. Sylvia Cranmer McLaughlin, one of the nation's preeminent environmentalists and founder of the Save San Francisco Bay movement, unveiled a plaque with a portrait of her husband, Homestake chairman emeritus Donald Hamilton McLaughlin, who had died on December 31, 1984, at their home in Berkeley; the mine was named for him. Homestake executives Harry Conger and Jack Thompson spoke, as well as Harold Moskowite from Napa County, Walter Wilcox from Lake County, and Twyla Thompson from Yolo County.

Following this ceremony there were two days of public events. A big tent was erected at the core shed to display a plexiglass column eight feet tall containing the 327 permit documents, as well as the model of the processing plant, and many pictures and diagrams. Krauss recalled,

> The first day we invited all the government officials and all of the construction engineers and technical consultants and folks that had a direct hand in the project.... The second day we had the public invited...somewhere around 7,000 people showed up. We had the whole meadow area north of the core shed opened up as a parking lot, and we still had the thirty-two buses that we had used for

employee transportation during the construction period, so we hosed all those out and put them on the line. They were all lined up along the slurryline road, and we had everybody that could speak English on our staff and had any kind of social skills at all to be tour guides and had one in every bus.

We just ran them in a circle. They would stop by the tent and pick up a load of people, and they would go out to the mine and drive around the mine and drive back up through the process plant and come back down the slurry line and unload people. We had refreshments and brochures and newsletters and yellow hats. It was a very, very well received celebration in the community.... And it... was reported in all the newspapers. There were people from Solano County and Sacramento.... It...set the tone for our presence in the three-county area from that point forward. It was clearly an acknowledgment from the company to the community that we were thankful for being welcomed into the community. (Krauss interview, KMD vol. 8, 355, 356, 357)

In spite of all the careful planning, in November, two months later, there was a surprise, as Krauss recalled:

[I]n the midst of mining one day, the Napa County Sheriff's Department and three different squad cars...came roaring into the mine pit with their lights flashing, and...they blocked the equipment and asked for the person in charge...Steve Drake, who was the mine superintendent.... [T]hey required the mining to cease.... Homestake and its attorneys would meet with the sheriff or the district attorney at eight o'clock the next morning to resolve whatever problem there was.

As it turned out, there was a five-dollar fee for an annual permit for handling blasting materials that had to be obtained on an annual basis from the county sheriff. The mine manager or the blasting crew normally had obtained that permit on a regular basis, and they had in fact gotten it the previous year, and they had failed to get it renewed. And so the sheriff decided to exercise his prerogative and shut us down.

But Denny...and Jack went down and met with the sheriff the next day. My recollection of their story is that the sheriff said, "Boy, you must have had to scrape the bottom of the barrel to invite 1,200

> people to that party and leave me off the list." . . . So that apparently was the genesis of his disappointment. It was clearly a clerical oversight. . . . [W]e went through all of the lists of people that had participated, county assessors and treasurers and supervisors and planning commissioners and public health officers—and certainly we intended [to invite him]; it was just a clerical oversight that he didn't specifically personally get an invitation, although there were also public invitations inviting anybody in the surrounding three counties to participate, so he wasn't excluded, but he felt slighted. (Krauss interview, KMD vol. 8, 305, 306)

John Drummond, Lake County Schools attorney, recalled the September event favorably:

> [A]t the time the gold mine had their grand opening, everybody . . . was invited to a giant barbecue. . . . [E]verybody was given Homestake hats. . . . I remember I got a hat. And thanked for their participation. And we were really left with the feeling that this was "our" mine, that this was Lake County's mine. And within a year after the mine opened, Lake County was hit with some terrible summer . . . brush fires that threatened to wipe out entire neighborhoods . . . and . . . without even being asked, Homestake immediately put its equipment on those fires. . . . And that kept community support for the gold mine. (Drummond interview, KMD vol. 2, 274)

Homestake also paid for the purchase of an extra fire engine.

The start-up was concluded, and the Manhattan project was now the McLaughlin Mine, to be developed.

CHAPTER 18

DEVELOPING THE MINE

McLaughlin has few, if any, rivals in the world.

—Dean Enderlin, Homestake geologist

In 1960 Nathaniel Herz, chief metallurgist at the Homestake Mine, wrote:

> Currently, the world's annual gold production is . . . about 1,000 tons. It is estimated that the entire production of all time is about 2 billion ounces, or nearly 70,000 tons, which would make a cube less than 50 feet on a side. (Herz 1966, 73)

Between 1985 and 2002 Homestake's McLaughlin Mine added about 3.4 million ounces of gold to this total. In May 1982, when Jack Thompson's feasibility study for the project was presented to the Homestake board for approval of the expenditure, he estimated the cost to be "a couple of hundred million bucks" for what would be the McLaughlin Mine (Thompson interview, KMD vol. 7, 41). In the end, as the company chairman Harry Conger recalled,

> This was a state-of-the-art mine, and it truly was, not only technologically but environmentally. Because of the way the mine was set up, greenfield mine start to finish, was really a mark of achievement for the people that did the mining. (Conger 2001, 160)

> The mine turned out to be a tremendous technical success, but financially, we just a little more than get our money back. . . . [F]rom '85, with the exception of about five months, we never saw $400 gold. . . .

> But here we were committed, the equipment was ordered, we were building the plant.... It wound up costing about $280 million to construct the facility, including the property acquisition. (Conger interview, KMD vol. 2, 188, 190)

THINKING IT WAS "LIKE THE WILD WEST"

The development of the mine changed lives in the neighboring communities. Marion Onstad, a Morgan Valley neighbor, was one of the first to be hired at the Manhattan project. In 1964 she and her husband bought ranch property there and built a log cabin from a kit. They wanted to escape from the high-tech world of Silicon Valley, where she was a secretary. In 1980 they moved to Lake County to reside, "thinking it 'was like the Wild West'" (Onstad interview, KMD vol. 6, 62).

Then, Homestake announced its gold discovery, literally next door. Soon after that, when she learned that Davy-McKee was seeking a secretary, she applied for a job and hurried to finish canning tomatoes before starting to work in a trailer equipped with a desk, a microwave telephone, and a fly swatter. When the project was handed over to Homestake, she continued with them, becoming secretary first to the mill manager, Joe Young, and later to resident manager Jack Thompson and his successor Ronald Parker. In this capacity, she organized workshops for better clerical personnel relations. After she retired in 1995, she undertook to run the mine tours for Homestake. She published a coloring book for children that features Rodney the Roadrunner and tells how miners respect the earth. The Onstads had a lease arrangement with Homestake and continued to graze cattle on the property around the office buildings. She welcomed the changes to Morgan Valley: paved road, telephone, and electricity.

EVERYONE WAS JUST ASTONISHED AND THRILLED

Bonny and Ross Hanchett owned and edited the *Clear Lake Observer*, and their five children were all involved in one way or another with the enterprise. The newspaper covered news of the entire county before and during the mine's operation, as Bonny recalled:

> [W]hen the news broke about the gold mine...everyone was just astonished and thrilled. It was very exciting for everybody.... We did many articles about it. We went out and took pictures. I remember one picture we took. It was this huge earth mover, and this

> little tiny woman was driving it, and we got a picture of her standing by one of the wheels, and she looked like a little doll in front of that. It was quite a news item to think that a small woman like that would be driving such a huge machine. (Hanchett interview, KMD vol. 4, 85)

Her son-in-law was one of the new hires:

> He was in heavy construction. He's an earth mover. He doesn't operate the machines, but he does the grading and he does all the surveying and the mathematics and everything to show them where to dig and so on. That's what he does for a living. He's been in it for years and years. (Hanchett interview, KMD vol. 4, 87)

Roberta Hanchett Lyons was reporter for the family-owned newspaper, and she recalls an early visit to the mine:

> I went out a couple of times and did some stories, and he [Ray Krauss] showed us how they blasted out the tunnels and showed us around.... They were gating the old adits to protect the Townsends' long-eared...bats...I interviewed the bat people [the scientists]. Dixie Pierson, the bat biologist. Bill Rainey...her companion, fellow scientist.... [There was] something they would set in the adit that would register the bats flying by, and then they would download it on their computer. So I thought that was pretty interesting. (Lyons interview, KMD vol. 5, 122)

TOWNSEND'S BIG-EARED BATS

For Ray Krauss, the project's environmental manager, this unforeseen situation was more than just interesting; it was a way to gain support in the scientific community and the general public. Townsend's big-eared bats are rare, and the successful—and well-documented—relocating of the bat colony was a heartwarming story: a big corporation going out of its way to save endangered creatures.

> ***BATS* Magazine**
> Volume 7, Issue 1, Spring 1989
> **Help for Townsend's Big-Eared Bats in California**
> By Pierson, Elizabeth D.
> When a historic bat roost disappeared, a gold mining company came to the rescue—

> ...the McLaughlin Mine, owned and operated by the Homestake Mining Company.
>
> ...in early June 1988, McLaughlin's staff environmental engineer, Dolora Koontz, biologist Bill Rainey and I were walking toward an old mine tunnel. Thanks to conservation efforts by Homestake, the tunnel is now the best protected maternity roost for one of California's most endangered bats, the western subspecies of Townsend's big-eared bat (Plecotus townsendii townsendii). A recently constructed, formidable gate protects the colony from human disturbance, and automatic monitoring equipment records bat activity in and out of the roost.... The formerly abandoned mine that had harbored the bats...had been removed by the mining operation.... the Plecotus colony, only one third its former size...[was moved to] an abandoned mine tunnel....
>
> What distinguishes the McLaughlin Mine is their firm commitment to wildlife conservation and their sincere efforts to provide enduring protection for their resident bats.... Ray Krauss, McLaughlin's pioneering Environmental Manager...[put] the mining operation on hold until measures had been taken to ensure the colony's safety.... Ray Krauss...said to me, "We are taking from this earth, and I want to give back." As a result, two very important Plecotus colonies have been saved, and a precedent has been set which should have far reaching consequences for bat conservation.
>
> Elizabeth D. Pierson received her doctorate from UC Berkeley in 1986. In addition to her research on bats in California, she is currently involved in projects on the ecology, conservation, and evolution of bats in New Zealand and the South Pacific. (Pierson 1989, n.p.)

The formidable gate that protected the bats was one example of the McLaughlin employees' concern for environmental protection, as Ray Krauss recalled:

> When we discovered the bat population in the workings and began efforts to figure out how to relocate them, a lot of people volunteered time, people from the geology department and the environmental department, and when we started fabricating a gate to protect their new habitat, the maintenance crews got involved and excited about helping on that.
>
> There really was a spirit of participation and environmental awareness from everybody.... For a while we had a program with

> the bird rescue center to release recovered raptors on the site, and we had a hack box up on top of the hill.... [Y]ou had to feed these raptors in the hack box...for two weeks or a month before you released them.... They [half a dozen employees] started picking up roadkill on the way to work in the morning, and they would stop by Marion's [Onstad] house and chop it up into pieces and provide it as food for these raptors. (Krauss interview, KMD vol. 8, 365–68)

SAMPLES TO SEE HOW WELL WE'RE DOING

Dolora Koontz, one of the environmental engineers who worked to rescue and relocate the colony of bats, had a historical connection with Homestake: her grandfather, much earlier, had worked at the Homestake Mine in Lead, South Dakota. In 1984 she and her husband, Gary, both biologists, were among the several Homestake personnel from other locations who transferred to the new project and fit in to the community. At the Pitch uranium mine in Colorado, Gary was chief mine surveyor. They came to Lake County in April 1984, and rented a home in Riviera West, on Clear Lake between Kelseyville and Lower Lake. Gary went to work right away as surveyor, "doing a lot of claim-line work, crawling through the chamise [Big Sage Brush or Chamizo Blanco, Artemisia tridentate] day after day" (Koontz interview, KMD vol. 4, 238). Dolora was first a technician in the metallurgical lab, doing sample analysis that directed operations at the pit:

> Everything...was basically an acid digestion type of procedure.... They would collect samples throughout the process at various points to see how well we're doing on recoveries. A lot of it was blast-hole or drill-hole analysis for gold assays. All that information goes back down to the mine so that they can then go out and resurvey the ore blocks. (Koontz interview, KMD vol. 4, 239)

She had been a forestry technician running the Dutch elm disease control program in Jefferson County, Colorado, and she used this valuable experience later, in the environmental protection of the McLaughlin project. In the summer of 1988, she was promoted to environmental engineer and in December 1995 to senior environmental engineer, with a wide range of responsibility, as she describes:

> I do water quality, geotechnical, wildlife, reclamation, some of the permitting and interaction with operations personnel on all those

> issues.... We have creek sampling...at...twelve stream stations... [w]e run about twenty-four different analyses on each sample site... [for] metals as well as general chemistry....
>
> [W]e also have thirty groundwater wells...twelve of them... surround the tailings impoundment so that we could detect any... seepage.... There are nine...observation wells which are drilled into the tailings dam embankment and allow us to detect...seepage.... [A]t the minesite and waste rock area we have nine wells and three underdrains that we monitor.... [T]hen on top of that we have our discharge sampling in compliance with our NPDES [National Pollution Discharge Elimination System] permit...that's issued by the state water board.
>
> Everything is done according to a protocol: you have proper sampling techniques so you don't contaminate your samples...fill out chain-of-custody forms to ship it off to the lab.... [E]verything is done with a record to back it up, and...all the data is reported to... all of our...agencies for review. We incorporate everything into one report. (Koontz interview, KMD vol. 4, 243, 244, 245, 262, 265)

Her skills would continue to make valuable contributions throughout the production phase of the McLaughlin Mine and also the reclamation.

A MECHANICAL ENGINEER PUMPS STUFF

Another transfer employee was Patrick Purtell, who worked at Homestake operations in Grants, New Mexico, and in Creede, Colorado, before he joined the McLaughlin project in 1985 as maintenance engineer. He later became manager, serving until 1997 and through the shutdown period. At first, he and John Turney, the metallurgist, shared a rented house in Lower Lake, and Turney describes him as "an engineer who really enjoys being an engineer; he just eats and sleeps with the problems until they are solved" (Purtell 1999, iii).

Purtell served for two years in the Army and saw battle in Vietnam before he graduated from the University of New Mexico as a mechanical engineer. Then he worked at the Kennecott's Chino copper mine in Arizona and at a Union Carbide chemical plant in Texas. At the McLaughlin Mine, he was not overawed by the problems:

> A mechanical engineer pumps stuff over here, and pumps stuff over there. We don't care what's being pumped. It's just pumped. And

> things turn. And you fix them, and you do this, that, and the other. That's all there is to it. No big deal. (Purtell 1999, xiii)

At a traditional gold mine, the ore is taken from underground or an open pit, first crushed, then ground, passed through a processing plant, and then to a refinery that produces a gold bar. Ideally, all this is adjacent to the mine and on a hillside so that gravity takes care of the flow. In the case of the McLaughlin Mine, because of the restrictions of environmental protection, it was more complicated.

Selecting the site for the discharge from the processing plant, called tailings, was the crucial decision. After painstaking search, the basin was finally selected in Lake County that had a layer of impermeable soil at its bottom, an advantage that outweighed its location, more than five miles from the mine pit. The next decision was the location of the processing plant: by the mine, or by the tailings pond? Again, environmental protection was a factor, and it was placed near the pond. The ore would be crushed and ground at the pit, mixed with water, and the inert slurry pumped five miles, avoiding the risk of a contaminating spill if the tailings were pumped five miles. At the mine, topsoil that was stripped off had to be stockpiled for replacement after the mine closed. Waste rock from the mining had to be piled separately.

All this resulted in employment and tax benefits for Lake County. The fourteen miles of road from Lower Lake had to be paved and strengthened to accommodate heavier loads. Water was needed for keeping down the dust, for compacting, and for making concrete. It was purchased from the Lower Lake Water District, and the water main in the town was upgraded to full public utility standards, thanks to the $160,000 compensation from Homestake.

A 115-KV power line was installed from Lower Lake to the mine, and a smaller line was put in underneath, providing power for the first time for residents along the road. At first the company used microwave telephone, and later installed a fiber-optics line that provided landline service to the neighbors as well.

Purtell recalled meeting one early challenge:

> So I had to install a twelve-inch pipeline for slurry being pumped from the process area to the grinding circuit.... Five-mile line.... It's buried all the way.... I started noticing pipe corrosion...showing in areas that you shouldn't have seen it, or pipe was just thinning up,

> and it wasn't due to erosion. I was concerned. I started seeing it at grinding, too. I mentioned it at one managers' meeting, like nine o'clock in the morning.... [I] realized that we have a problem that we're going to have to address somewhere along the line. At two o'clock that afternoon we got a call on the radio that there was slurry blowing up in the air about two miles from grinding. Shut the pipeline down coming up here and dug it up, and they had a hole corroded in it. So we cut out a piece of the pipe, and the whole bottom of the pipe, as far as you could see, had spots, pits, corroded in it from the ferro-thyobacillus.... It's the bacteria that...will actually digest iron. Likes to eat iron.... It will also eat pipelines. (Purtell 1999, 49–50)

He solved the problem by lining the pipes with plastic, and the operation continued.

IF THEY WERE TOLD TO MEASURE A CERTAIN AMOUNT, THEY WOULD MEASURE THAT AMOUNT

In many parts of the world, it had long been considered bad luck for a woman to enter a mine, even as a visitor, let alone a worker, so Homestake broke with tradition when it hired several women to work in the mine. One of the first was Della Conner Underwood. She and her husband Curt were cattle ranchers, just down the road from the mine toward Knoxville. Her family had been ranchers in that area since the 1860s, more than a century earlier. When she was born, her parents were living in a house at the Knoxville Mine. She went to the Knoxville school and graduated from Napa High School in 1961; a few months later she married Curt Underwood, and they began their life as ranchers together.

When they heard that the mine was hiring workers, Curt applied for a job as truck driver, but failed to pass the physical exam because of earlier back surgery. Della recalls her own hiring:

> I kind of got to know Marion Onstad, who was the secretary up here.... [S]he said, "Turn in your application, and if I get a chance, I'll put it on top of the pile."...So, yes, that's how I got hired.... [It] was in August of '86. I think they had already been working... almost a year, before I got on.... I applied as a laborer, and the first job I got was on the blasting crew.
>
> [T]he head blaster was Shorty Watson...and he wanted to hire all women because he felt like if they were told to measure a certain

> amount of something, they would measure that amount.... [T]here was Liz Thomas. And Bill Wilder's daughter, Kelly [Wilder].... And Linda [Lucientes]. [And] Kathy Montie.... We were the crew.
>
> ...On the blasting crew, there's just certain things, steps that you did each day.... You know, they had a hole...the drillers drilled the hole.... [Shorty] knew how hard the ground was...so he'd know that this area's holes would have to have so much powder, and that area so much powder, and we just basically done the same thing every day. It was kind of exciting...the first few holes. (Underwood interview, KMD vol. 7, 311, 312, 313, 314)

IT WAS HARD, BACK-BREAKING WORK

The excitement wore off and it became a grueling routine:

> You'd put a booster down the hole, you put the e-cord on the booster, and you put a certain amount of powder in the hole, and then you took a little back-breaker and hoed all the shavings back into—what they call "stemming," stemming the hole, and you just had to put all that stuff back in the hole.... And then after you got all the holes loaded, you'd have to squat down and tie them all in. There was a certain way you'd tie the lead-in cord to the blasting cords.... It was hard work. I mean, it was hard, back-breaking work.
>
> Five days a week [in any kind of weather]. And we put our little rain suits on, and it would get muddy, and if you're out there hoeing that shavings back in, you know, you'd have mud all over you.
>
> ...[T]hree o'clock in the afternoon is when they usually blast. And then we set off the blast at three o'clock, and the next day we'd go to it again.... We were setting off the maximum amount per day, which would be right around 300 holes. (Underwood interview, KMD vol. 7, 313, 314)

I WAS A LITTLE LEERY

After working four months on the blasting crew, Della went to work in the engineering department:

> I was a little leery because, you know,...I thought men were always—you're supposed to take orders from them.... [M]y dad and mom were always, like, equal. And my husband always treats me

equal. But you're kind of a little leery going to work with all these guys, you know?... But they're really good... to work with....

Well, basically... we are engineering technicians. And the engineering department do all the designing of the pit area, and they... [have] to know where the gold is.... So we have to know exactly how much dirt is being taken out of the mine every month, so we do a complete survey of all the areas that have been mined each month and figure the volume on the tonnage for that.

...And the drillers, when they drill, they leave a sample bag by each hole. One of our jobs is to go out, give that bag a number, make sure that's recorded on the map, deliver it to assays.... All the samples have to be logged and mapped so... when you get your assays back, you know where to mine them.... And then when the assay comes back they have ore control people, and... those guys are geologists.... [T]he ore control person, he draws up what we call blocks on his computer, ... where these holes were.

...We go out there, and we shoot in those corners for each one of those blocks and tape off that area... and then we put pin flags on each one of those blocks that are a different color, and that tells the operators then when they're mucking that this block here is high-grade; this block has nothing in it.... I've learned to operate the surveying instruments. You know, do what we call "run the rod."... [A] rod person has to hold the rod for the person who's running the survey instrument.... So... it's pretty physical.... You're not only on the computers; you're out there... putting ore blocks... and you're climbing up and over, and [you're] the person with the rod and the person with the flagging and the stakes. (Underwood interview, KMD vol. 7, 315, 316, 317)

Computers have made the work easier:

[W]hen you used to survey, you had to have a person take notes as to the angles and things you were doing. Well, now you just hook up your little pack and put some information into it and punch a few buttons, and it records all that. You bring it in and you plug it into your computer.... [A]t the end of each month, when we do the volume survey, it puts all those points that you've surveyed right on a map, so you know the exact elevation and distance and all the information you need....

> I've been responsible for logging samples and delivering them to the lab. I've learned invaluable stuff on the computers that I had never had any experience with before I started here. (Underwood interview, KMD vol. 7, n.p., 316)

I THINK IT'S TIME TO RETHINK OUR SCALE HERE

In the days of California's first gold rush, some fortunes were made by those who supplied goods to the miners. The classic example is Levi Strauss and his sturdy copper-riveted denim pants. The McLaughlin Mine, during its life, brought prosperity to local merchants; one of them was James Jonas, the second-generation owner of Lake County's principal bulk fuel plant, in Lower Lake. In 1997 he recalled how the mine changed his business:

> First, you have to understand we're sitting here and that mine is about twenty miles up this road. The road that exists today that Homestake built is like a freeway. It's probably one of the best roads in Lake County even after all the traffic has gone over it in the last ten or twelve years. In those days, it was a dirt trail; I mean it would take you half a day to get out there, almost, particularly in the wintertime. So we were the natural ones to supply that....
>
> [S]omewhere around the early eighties Homestake started in there and it was around '83, '84 that they really got going.... I was already there with the One Shot Mine and so when they first started going out there to do their exploration, and their core drilling and what not, it was kind of a natural; I was there. Wilder of course tipped me off to what was going on.... As soon as I knew that something was going down, I got a hold of Homestake and Davy-McKee and everybody that I could find out was bidding the job,...from all over the country, I sent them letters.
>
> ...The first purchasing agent was Wayne Bean, I think was his name,...and later, a fellow by the name of Dwight Martin came along.... These are not people who will probably turn up in the history books....
>
> The first day they called me and they said, "Bring some fuel out here. You've got to fuel a couple scrapers." I said, "Okay." I had this little thirteen-hundred-gallon truck out here and I put about two or three hundred gallons, which I thought would be plenty. I got out there and I looked up at the tank and it was fifteen feet up there, and

it's a five-hundred-gallon tank hanging on the side of this immense scraper that towers higher than my head.... I said, "I think it's time to rethink our scale here a little bit."

...Well, they first started core drilling out there, Homestake did, and we delivered to their core drilling operations, diesel fuel and gasoline; we had tanks out there. Then they brought in the heavy stuff; that was totally amazing.... The stuff they started bringing in, the size of the equipment, was immense compared to—we're used to a D-8 tractor and a little paddle-wheel scraper or something to move a little dirt.

...One time we were hauling—. I think we got up to twenty to thirty thousand gallons of diesel fuel a day. We were taking twenty, thirty, forty barrels of lubricants out there a week. It was just something.... We probably went from two, or three, or four people to twelve to fifteen in a pretty rapid period of time. (Jonas interview, KMD vol. 4, 173, 200, 202, 203, 208, 209)

THIS REGION WAS SWARMING WITH CREWS OF GEOLOGISTS

In July 1985 Homestake hired a young geologist, Dean Andrew Enderlin, a fourth-generation resident of Napa County, who tells a classic family story of his great-grandfather coming to California:

> Supposedly he was high-ranking...in the Kaiser's personal guard.... [W]e know that he swore that he wanted no part in the military.... In the 1880s in Germany you had Bismarck and the Kaiser becoming increasingly warlike.... [W]hen the ship was passing across the Atlantic, he made sure to ask the captain at what point they reached the middle of the Atlantic. At that point he ceremoniously threw his medals overboard and closed the door on all the past in Germany. (D. Enderlin interview, KMD vol. 8, n.p.)

After a brief stay in San Francisco, in the 1890s the family moved to Calistoga, acquiring land at the head of the Napa Valley, where Dean's father was born, and where he was born in 1961. His interest in gold began early; as a boy he panned for gold at a relative's ranch near San Andreas in Calaveras County. In 1985 he graduated from Sonoma State University with a bachelor of science degree in geology. This led to his work as a guide on a neighbor's property, as he recalled:

> After Homestake had made its announcement, this region was swarming with crews of geologists. It was kind of a standard practice for me to tour geological reconnaissance groups....
>
> Over the hill, to the north of our property, there are a series of old patented mining claims. Patented, meaning purchased from the government. And this elderly couple [Maurice and Enid Williams] lived on these claims.... She was the granddaughter of...Robert F. Grigsby, who was the founder of the Palisade Mine, the adjacent property in the 1880s.... [It] was a gold and silver mine, primarily silver.... Their property consisted of a series of mining claims that Grigsby had patented, but never developed. He had always told them that these claims would provide for the family in later years, that there was plenty of silver there.... They were elderly and they really couldn't get around very well to show all their mining claims. It was Maurice's dream to see the Palisade Mine go back into production. I would accommodate every group of geologists that came to their property by giving them the official tour around the site. Not just their property, but also...the Palisade Mine. (D. Enderlin interview, KMD vol. 8, 8, 18)

In July 1985 a group of Homestake geologists came, and at the end of the tour Norman Lehrman offered Enderlin a job doing reconnaissance on a potential mill site—not searching for gold, but to make sure that there was no gold there. He arranged to live with a distant cousin, Elmer Enderlin, in Lower Lake.

Elmer worked at fifty-eight mines in his long life and was a published poet (E. Enderlin interview, KMD vol. 3).

> [H]is house...on Highway 53 in Lower Lake, was directly across from Homestake's Park-n-Ride. At that...time Homestake was running buses for all the shifts. It was a simple matter for me to wake up in the morning, have breakfast, and run out the door and across the highway and hop in the Homestake bus.... The bus would bring us up the hill and return us, so I had a good deal going there.... [They] gave me a Honda ATC, a three-wheel all-terrain cycle. This had been used by the Environmental Department.... [M]y primary task, was to construct a geologic map of the area west of the mine, and to collect samples.... I took hundreds of samples out there. (D. Enderlin interview, KMD vol. 8, 22, 23)

Two years later, he began to work in the mine pit, in a job that bridged geology and mining:

> [T]he one thing that has been a rule from the 1860s to the present, is: you don't run waste rock through your mill. Some things never change. Dilution is your enemy.... [A] mining engineer...must design the pit mining schedule so that you're never in a situation where all you've got ahead of you is waste rock. A mill can be a very hungry thing, and a very expensive thing to shut down.... We... analyze all the cuttings from every blast hole [to] give us a good estimate in the field of where the highest grades lie, but you can still, without care in mining...dilute the ore going to the mill. (D. Enderlin interview, KMD vol. 8, 42, 51)

He knew about the weather in that part of Napa County:

> [T]hat particular part of the region is above the fogs that come into the Bay Area, so...the summers are very...hot. And...because of the elevation, which is around 2,000 to 2,300 feet, the winters are a bit cold, and you can get some snow and...chilly conditions. So you get extremes. You don't get huge amounts of snow, but certainly enough to be a cold, wet snow.... So wintertime was...not a pleasant time in the pit whether you were in equipment or out in it. (D. Enderlin interview, KMD vol. 8, 70)

In 2001, when he was manager of reclamation for the mine, Enderlin looked back to his time as ore control geologist in the mine:

> No matter how deep we mined into the rock, everything had to be drilled, everything had to be blasted, but all that rock would just turn to mud in the middle of winter. The heavy equipment just churns it up, and so that made for a very sloppy, miserable time of year, but production went on.... We had remarkably good communication and good rapport between the gold extraction people and the mine people, even though we referred to each other as the mill guys and the mine guys....
>
> The ore control geologist in those days...[was] on the ground walking around in the midst of haul-truck traffic. You wore your orange vest, and you made it very clear where you would be.
>
> There were no radios at that time; in later years we installed radios in all of our equipment. The ore control geologist had a

hand-held radio like a walkie-talkie to communicate with other staff in the truck shop or in the administrative offices, but none of the equipment operators had radios, so we had this system of hand signals, and it was quite an elaborate system... it was specific to ore control. There were signals for ore, signals for waste, signals for hard rock, signals for soft rock, signals for the toe of a bench, signals for the crest, just an elaborate hand system. (D. Enderlin interview, KMD vol. 8, 46, 47, 70, 73)

A system of colored plastic flags defined ore content:

It was more than just determining how much gold was in the rock. Because the autoclaves that were very sensitive to sulfide content and carbon content, as well as the gold grade, we had a fairly elaborate system of defining different types of ore and staking those with the appropriate color in the field.... We used a... rainbow-colored spectrum, with hotter colors like pink and red indicating higher-grade ores, and cooler colors like green and blue indicating lower-grade ores. White always denoted waste rock, so wherever you saw a white flag you knew that you were in barren ground.

The flags were placed by the survey crew after every blast pattern, and their placement was very precisely controlled.... The... patches of different color were bounded by stakes; about three-foot-tall lath would be pounded into the ground, and... a real thin... ribbon would be strung between posts to create a geometric shape. Not real bulky ribbon, since you went through miles of the stuff; we had to carry it with us out into the broken rock. What the equipment operator would see in the field would be... this sea of different colored flags, the stakes in the ground, the... ribbon strung between the stakes to outline the blocks, and then they would be given a map showing the outlines of all these ore blocks ahead of them.

One of the functions of the ore control geologist was to update that map to always show the current mining face. We would mine in pit benches twenty feet at a time; a vertical 20-foot cut... because that was the optimal height for the [Caterpillar 992C] loaders.... The buckets were capable of scooping 20 tons of rock per bucket load, and they would fill the Caterpillar triple sevens, which would haul on the average around 80 tons of rock. In later years when we purchased hydraulic shovels, we added sideboards to the triple sevens,

> the haul trucks, and then they could carry up to 100 tons if they were heavily loaded. (D. Enderlin interview, KMD vol. 8, 47, 49, 50)

He explained the significance of the McLaughlin deposit:

> [McLaughlin] is probably the world's best example of a perfectly preserved cap of an epithermal system.... You never make new gold. It's only recycling, you might say, in nature, and the hot springs simply created the conditions where gold could dissolve by hot water deep in the earth and be transported to some central location, like a gigantic concentrating machine.... The McLaughlin deposit... still had preserved at the surface the hot springs sinter. These were the ancient terraces where the water...flowed upwards through these fractures in the rock...and flowed out onto the landscape.... [The] sinter...was a surface deposit...formed by the hot water as it emerged.... [It] gives you a datum.... this was the original surface one million years ago. That made McLaughlin very special, and one that has few, if any, rivals in the world. (D. Enderlin interview, KMD vol. 8, 26, 30)

IT'S TIME TO DOUBLE THE THROUGHPUT OF THIS PLANT

Ten years after Don Gustafson first walked onto the site, the McLaughlin Mine was an ever-growing pit, and the autoclaves were processing ore and producing gold. When Homestake's board first approved the project, the resource was estimated at 1 million ounces of gold, making it a world-class mine, and the price had just declined from a peak of $600 an ounce. A decade later, the reserve estimate had expanded to 3 million ounces, with an expected recovery of 2.8 million. Costs had escalated, however, and the price had not. Jack Thompson, now in the San Francisco office, decided it was time to double the throughput.

Ronald Parker was the new manager at the mine, given this assignment. He was born in Boss, Missouri, in 1950, and graduated from the University of Missouri as a mechanical engineer. In 1986 he was working for AMAX at the Buick, Missouri, lead smelter, and was recruited by Homestake to move to the McLaughlin Mine, which he called "the newest, most exciting mine in the United States" (Parker interview, KMD vol. 6, 159). He started out as chief plant engineer; two years later he became general manager, the first mechanical engineer to be promoted to general manager. His wife Merrily was a schoolteacher, and when they moved to California, she immediately began to teach in the Cobb school in the Middletown district. Ron Parker recalled his first impression of the McLaughlin Mine:

> [T]he things that you saw immediately was the attention that we paid to the environment.... Phil Hocker is with the Mineral Policy Center, a fairly radical group which was against mining.... [He] and his group came out at least three times during my reign as general manager.... several of Phil's articles stated that the only good mine in his opinion was the McLaughlin Mine. (Parker interview, KMD vol. 6, 163)

Parker gave credit to Raymond Krauss, the environmental manager, for easing the permit process for the new mandate:

> [T]he construction... [of the new circuit] took a whole new set of permits... because we doubled the through-put. Now, we were able to get those permits through the county in less than a year.... Which was a remarkable achievement by Ray Krauss.... [It] said the way that McLaughlin operated up to that time. We had a track record of doing what we said we will do. (Parker interview, KMD vol. 6, 172)

Krauss also recalled the process:

> [T]he hours had been long, a lot of weekends, a lot of twelve- and fourteen-hour days for a long time.... [Start-up] was completed and we had begun to regularize all of these monitoring and... reporting documentation programs, it's like seeing the light at the end of the tunnel... and then Jack [Thompson] came along and said, "Well, now it's time to double the throughput of this plant."... And so again we were in the process of working with Davy-McKee in designing, completing the engineering that we needed to describe the new project, the expanded project and its consequences.
>
> [W]hen we went through the permitting process we were so favorably received in the counties that there was no opposition in any of the three counties to the permits that allowed us to double the throughput. (Krauss interview, KMD vol. 8, 368, 369)

Parker's team improved the operation of the autoclaves:

> [W]e had quite a team.... On that team was Pat Purtell, also a mechanical engineer.... It was a remarkable team... that could organize their thoughts, organize the logic, and tackle the problems

in the right order.... The feasibility studies assumed that we could get 85 percent availability and that we could treat 300 tons per day through them. I think when I got there in the summer of 1986 we were in the mid-60s on the autoclaves, so it was running about two-thirds of the time, which wasn't bad. It was a brand-new device, never used before in the gold industry, and really it was remarkable achievement, ...to have it to 65 percent. By the time I left [in 1994]... we are achieving around 92 percent availability, so only down about 8 percent of the time. (Parker interview, KMD vol. 6, 161, 162)

Dean Enderlin recalled what doubling the capacity of the mill in 1988 meant for him as the ore control geologist working in the mine pit:

When we built the second mill and sped up the whole process, we also accelerated the mining rate.... [A]s a consequence of changing one key component in the whole system, the whole network of mine planning issues was adjusted accordingly....

So the ore control geologist's function was to...select given ore blocks from different areas in the pit, and see that they were delivered to the lot pad.... [A] given lot was usually 3,000 tons or up to about 10,000. The lot was a combination of multiple ore blocks... denoted by the colored pin flags.

...[T]here was a substantial tonnage of ore that had been naturally oxidized that didn't require the autoclave pretreatment to artificially oxidize the ore.... That was another one of our pin flag colors; blue for CN-ore; CN short for cyanide, which were set aside specifically to be run through the secondary circuit.... [B]y bypassing...the enormous expense of operation of the autoclaves...those ores were very economical to process. (D. Enderlin interview, KMD vol. 8, 56–57, 61, 62–63)

Ray Krauss praised the operation:

Metallurgically it was really quite an elegant process. The addition of the float [flotation] plant allowed us to develop a sulfide concentrate that could be added to the autoclave circuit as it was needed to optimize the operation of the pressure oxidation system. The autoclaves functioned best at a steady concentration...of around 4 percent or 4.2 percent [of sulfur]. Prior to the float plant, we had made great efforts to blend ores coming out of the pit to maintain that sulfur

> level, but it wasn't always possible to do so.... But with the addition of the float plant, we were able to create our own sulfide concentrate out of the low-grade ore.
>
> ...[I]f you look at the sequence of metallurgical solutions...it's really quite elegant. You start out with the autoclaves, you add the oxide circuit, you convert the oxide circuit to flotation and low-grade processing, and you end up with the low-grade ores processed through direct cyanidation, and it really optimized the recovery from that ore body. (Krauss interview, KMD vol. 8, 371)

John Turney, who helped to develop the autoclaves, once said,

> Just because we've got some ground and it's got some gold in it, it doesn't mean we have a mine. What makes the mine is, "Can we dig that material out of the ground? Can we process it, make a gold bar, and make a profit?" (Turney interview, KMD vol. 7, 276)

His rhetorical questions could now be answered at McLaughlin, yes and yes.

PART 5

CHAPTER 19

SHUTDOWN AND RECLAMATION

It will be a lake.

—Patrick Purtell

Patrick Purtell was maintenance engineer at the start-up of the McLaughlin Mine on January 10, 1985, when the first ton of ore was dumped in the crusher. He later became mill manager, and then, in 1994, was promoted to general manager, responsible for the shutdown phase and continuing reclamation and conversion to a nature reserve. He had prior experience; from 1968 to 1974 he was plant engineer for closing down Homestake's Ambrosia Lake uranium mining enterprise near Grants, New Mexico. Then he worked on closing Homestake's Bulldog Mine in Creede, Colorado. This shutdown was different, however; it was also a start-up in the processing plant and the continuation of reclamation efforts. Just as he had said that start-up was fuzzy, so was the shutdown.

The company set 1996 as the date when the flow of high-grade ore would end. The autoclaves, no longer necessary, would be opened up and cleaned to get rid of any accumulated scale. The agitators would be dropped back into place, and the equipment would be ready to sell. John Turney, the metallurgist, had overseen the autoclaves from the beginning, when he traveled the world to find the best way to recover gold from refractory ores. Now they were to be obsolete. He recalled,

> That will be not quite eleven years.... [You] know, the original concept was fifteen to twenty years at 3,000 tons a day, but six years ago we added that extra grinding train, and then mining at 6,000 tons a day, so it's not surprising that the life is condensed. And what with

> control of costs and a little bit more ore, we've been able to process more tons than was originally anticipated, so the three million ounces that were believed to be there were definitely there, and I think we're well over two million ounces out of there now already. (Turney interview, KMD vol. 7, 300)

After the autoclaves were taken down, the processing plant would treat by conventional cyanidation the low-grade ore that had been stockpiled. This meant adding cyanide tanks; since there might be more emissions of cyanide, Ray Krauss, environmental manager, had to revise all the permit applications for operations and air quality in Lake County. Thanks to his good relations with the county, and Homestake's record for compliance, the permits were soon granted unopposed.

IT'S A TOTALLY DIFFERENT MINE

In 1994 the mine pit neared its maximum projected size. Geologist Dean Enderlin had received notice that he would be laid off, but he was still in charge of ore control in the pit, and it was his busiest year so far. They had made an exciting find:

> The chief mine engineer...was Jim Whitlock. [He] had been down in the drift and had just inspected a recently blasted muckpile.... Jim came in with a little specimen of quartz that had some stibnite. Stibnite is an antimony mineral that's very common at McLaughlin. And in that stibnite were little dendrites of gold.... [H]e brought it in, and all excited.... He and I went back down there, and...on the... right rib of this particular cut, and here was about a nine-inch-long exposure of free-growing gold dendrites, standing out above the rock about a quarter of an inch or so. The coarsest free gold we've ever seen. Just a beautiful specimen. (D. Enderlin interview, KMD vol. 8, 67)

He put the specimen in the vault at the office, to be part of the permanent record of the mine.

Enderlin's crew was running three diamond drills at the surface, and also doing exploration in the pit; the target was a zone of concentrated high-grade ore. Just as they came to what he called "a pretty sweet pocket," they met "terrible sheared mudstone" (D. Enderlin interview, KMD vol. 8, 66). Pat Purtell was in charge and as always, put safety first in his decision:

> I'm the one that shut that pit down...we had four benches to go, and most of the gold was in the bottom two benches.... [We're] putting people at risk down in the hole, so we said, "That's enough."
>
> ...Thirty thousand ounces are still down at the bottom. (Purtell 1999, 74)

He predicted the future of the pit:

> We call this pit self-reclaiming, and already it's starting to do it by itself.... [When] this water starts to come up and it starts soaking the bottom of the west wall.... Nature is going to pull that wall down. The east wall that has got mudstone exposed, it's going to decrepitate, and...cave in,...until it finally reaches an equilibrium slope.... [It's] going to be a very deep pit. (Purtell 1999, 87–88)

In 1997 Purtell was already planning for the shutdown, six or seven years down the road. California law required an approved reclamation plan before mining began, so from the first days of operation, reclamation and mining were simultaneous. Over the years of operation, Homestake accrued funds, an amount for each ton of ore, to cover the eventual cost of reclamation. From the very beginning, scrapers took the top two or three feet of topsoil from all construction areas and placed it into stockpiles that were protected with straw bales to prevent erosion, and then reseeded and saved for use years later. The soil that was stripped from the Davis Creek reservoir site was shipped several miles to the mine site, to be used later.

The mine waste dumps—big piles of dirt and rocks—had been reseeded all along, and Purtell estimated that by the time the mine closed, there would be only two months of work left, final grading and smoothing, before they could plant trees there. If all went according to plan, they could hire students from UC Davis to do the planting. Drainage from the reclaimed dumps would still have to be monitored for a long time. The plan was to use a passive treatment system to bring the acid flows into a wetlands or a marsh, where bacteria would take care of the problem. Several small ponds were already there, with tules (reeds) flourishing, and trees on the banks that survived the heavy rains of 1995.

In 1997 shutdown and reclamation of the mine pit were hard to distinguish from each other:

> You send your drivers and loader operators and 'dozer operators out and start finishing reclamation work, looking as if we're finishing

> up the mine. So you're not making gold when you do that.... It's a totally different mine.
>
> [When] the turnover comes,...probably in 2003 or '04, we'll already have people actively working on projects, and there will be a much smoother turnover that way.
>
> So here we are, seven or eight years away, and we're already working on the shutdown that's going to happen in seven or eight years. (Purtell 1999, 75, 86)

Reclamation of the tailings pond also began at the start of mining operations, but plans were changed later, as Ray Krauss recalled:

> [Originally] we designed a traditional closure, which involved evaporating and dewatering the tailings and shaping them...and covering them with...two feet of soil...to turn it into a dry grassland.
>
> We have since revisited that whole question.... [T]hose tailings have been situated there for almost twenty years now without any evidence of release into the surrounding groundwaters or...surface waters...draining and capping the tailings is really no longer warranted.... [R]ather than reshaping it, we could reclaim it as a bowl, as a wetland, and ultimately save a lot of money by...coming up with a really interesting habitat that would be a mixture of grassland and wet areas,...that would support a whole variety of species.... So that's what we're working on. (Krauss interview, KMD vol. 8, 404, 405)

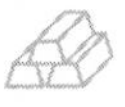

The shutdown affected all the employees. Purtell reluctantly complied with the decision made in San Francisco that everybody would be laid off at the end of June, and they could then volunteer to stay on. In the first round, thirty mine operators who had been hired as temporary employees were laid off; 180 more would follow. Purtell estimated that 110 to 120 employees would be needed to stay on until the end. Compliance with the Equal Employment Opportunity Commission (EEOC) standards required notifying an employee sixty days in advance of being laid off. Some of the salaried employees had a chance to go back to lower-rank jobs. As an added incentive, a stay bonus was offered. This acts like a bank account: an increasing amount is credited for each month that the employee stays, and if he stays until the last month, he gets the entire amount. Many workers wanted to stay over and finish up the reclamation work.

Well ahead of time, Homestake formed a transition committee made up of employees at all pay levels who served as a communications link and discussed what services they wanted Homestake to provide during the layoff period. Many wanted to stay in mining after leaving the McLaughlin. They had become used to making nearly $20 an hour, and there was a lot of activity in the mining industry, offering high-paying jobs. Homestake held a job fair at the Brick Hall in Lower Lake; eleven mining companies came to meet and interview workers.

There was a Job Training Partnership Act of 1982 office in Lower Lake to serve unemployed workers. When Purtell notified them of the layoffs to come, they received a federal grant to handle the load for providing counseling and stipends for retraining and financial support.

For Purtell, safety was always a primary concern:

> [It] was a period of turmoil, and it's hard to keep people focused on safety.... And we only had the six lost-time injuries; two true lost-time injuries during that period was pretty remarkable.... We've been proud of that one.... We just had MSHA [Mine Safety & Health Administration] in here just the other day, and he was bragging about what a great operation it was again. It seems that every inspector who comes in here decides that this is one of the better ones he's ever been in. That's a good source of pride, too. (Purtell 1999, xiv, 115)

JUST PUTTING THINGS TO REST

Dolora and Gary Koontz were looking forward to moving on when the mine closed. They both had degrees in biology from Western State College in Gunnison, Colorado, and first worked for Homestake at Pitch, Colorado. He was a mine surveyor; she was a technician in the engineering department. In 1984, along with Tim Janke and Roger and Pam Lucas, they transferred to the McLaughlin operation, and Gary began work as a surveyor checking out claim lines for the proposed mine. Dolora was hired in 1986 as a sample technician in the metallurgical lab. In 1996 he was chief mine surveyor, working for Dean Enderlin, and she was chief environmental engineer, working with Ray Krauss. They were foresighted and sold their house in Riviera West and moved to a rental so they could be ready, when their jobs ended, for an adventurous trip throughout Alaska, Western Canada, and the desert Southwest, before settling down in Fort Collins, Colorado.

Her responsibilities had expanded in the decade since she began as a sample technician. From 1995 she had been senior environmental engineer, doing a wide range of monitoring and reporting:

> We incorporate everything into one report.... You have your paper trail from step one on.... We have at this point thirteen years of data to draw on.... It's all there. (Koontz interview, KMD vol. 4, 244, 245, 265)

She was pleased with the success of the bat relocation, and anticipated a next step:

> [W]e will put in a hibernaculum [a hibernating site] for them.... [W]e're looking at...a creative use for the old haul truck tires. We're going to tie them all together to create a tunnel.... They're a good twelve feet tall...there's definitely plenty of space in there.... Dixie [Pierson] has helped us come up with a design to configure it so we can get different temperatures in different areas.... We're going to cover it all with dirt, put gates on the end, and hopefully the bats will use it.... It would be interesting to see if they would accept that type of a roost site. It would be a good use for old tires also. (Koontz interview, KMD vol. 4, 257, 258, 259)

The company kept her well informed about the shutdown.

> They've started...about a year ago...what they call the "Breakfast Club" meetings.... [P]eople feel like they're in the loop and being kept informed.... Don Field, our personnel manager, is starting to...help place people, to help them search for jobs, and...write résumés.... I think they're handling it well.
>
> [It is] interesting to see a close-down of an operation.... Just finalizing everything, making sure that the records are complete... and getting all the files in order,...ready for archive.... [J]ust putting things to rest. (Koontz interview, KMD vol. 4, 234, 272, 273, 274)

When Gary's mine surveying job ended and Dolora Koontz prepared to leave hers, Dean Enderlin welcomed the invitation from Ray Krauss, environmental manager, to replace her.

> [F]or post-mining issues, in addition to the routine scientific compliance issues,...there would also be a greater need to understand geochemistry,...those issues that would come up with closure: the stability of the mine pit as it begins to fill with water; the stability of the waste rock dumps; and things that start to move back into a geological realm again. (D. Enderlin interview, KMD vol. 8, 75)

He had already worked with Krauss when they had to comply with permit requirements. Like Krauss, he led tours of the mine site and gave talks to community and professional groups. Enderlin was also a local historian and concerned about preserving the mine's record.

> [M]ines always come to an end. But this is one that should be talked about for a long time.... [W]hat typically happens is, when the mine shuts down, all the paperwork is vandalized and lost. The miners are...gone. And there's no record of what happened. And enter the researcher who is tearing their hair out trying to piece together the story of a mine, with very few people around who know anything about it.
>
> So we're going to try very hard to not let that happen at McLaughlin.
>
> So the core collection fills in that void.... It was...a collection of bore samples from every part of the ore body at different depths, and precisely located, precisely surveyed so that if an interested geologist should wish to know what the rock looked like 500 feet below the surface at a given location, they could do that.
>
> ...We have a rock collection that we fondly refer to as Stonehenge,...some of the larger samples that we preserved and set aside.... Norm Lehrman and Jeff Wilson saw to it that key pieces of rock from different areas of the deposit be set aside and ultimately shipped to the core library area for display, just to represent the deposit. (D. Enderlin interview, KMD vol. 8, 86, 87)

They already had important visitors:

> NASA-Ames Research showed up at the door to look at samples of rock from our hot springs terraces that contained fossilized bacterial remains...Jack Farmer,...whose title is exobiologist, was involved in a worldwide program of collecting photographic evidence for primitive life on earth.... To think through...how they [landers on Mars] would go about sampling and looking for evidence of life....

[H]ere we are in Lake County and Napa County, our little back door, and we had Mars researchers, NASA researchers, showing up to photo document our collections because we had this unique scientific information. (D. Enderlin interview, KMD vol. 8, 86)

Looking back, Ray Krauss believed that the McLaughlin staff excelled in a number of areas:

> Certainly the skill of the process operators was incredible...to run an autoclave process that was novel, was unique. And they did it with great skill and success.... And safety.... [T]he maintenance group, in particular...Ivan Markland...and his engineers...Bilhartz and Brown...excelled in keeping [the autoclaves]...in operation.... Jim Fleming was the maintenance manager in the truck shop, with responsibility for maintaining the truck fleet and the loaders and all the heavy equipment. Again, it was an exemplary system.... Steve Smith was the safety manager.... [T]he safety record at McLaughlin was always among the best in the industry.... Knollie [Sells] had responsibility for all of the administrative functions,...all of the strategic planning...that kept the mine competitive even in the face of what seemed to be an ever-declining gold price. (Krauss interview, KMD vol. 8, 358, 359, 373)

He praised Jack Thompson's management style:

> Jack always was very encouraging to his employees. He kind of set the relationship as,...I hired you because I think you're the best person available to do this job. Now go do it. And I'm not going to get in your way.... [K]eep me informed."...
>
> It was a real flat organization. We had tremendous communication horizontally.... [I]f we had a problem between the environmental group and the mining group,...my environmental staff would talk to the foreman. He talked to the people in the pit.... And after the problem was solved, he would let me know, and I would let the mine manager know.... And we both let Jack know.... We didn't have to go up and back down.... [I]t was certainly central to what I have characterized as the McLaughlin culture. There was this sense of teamwork and mutual respect at all levels. (Krauss interview, KMD vol. 8, 360)

Phil Barnes transferred from the mine at Pitch, Colorado, and was the best environmental engineer Krauss had ever known. He compiled a procedural manual containing all of the monitoring requirements, sent it to all of the regulatory agencies, and they approved it. Eventually, the regulators had so much confidence in the monitoring that it could be reduced from monthly to semiannually. Phil also organized all of the monitoring results into a single annual report, sent each year to all the agencies so they could select the data relevant to their jurisdiction. Krauss recalled the impact of this report:

> This is a three-volume set that includes all of the chemical data from the analytical labs; all of the quality assurance data; all of the independently conducted engineering, geotechnical surveys and reviews; all of the aquatic ecology studies and the underlying data.... It goes to each of the about fifty agencies from whom we receive permits.... They joke about it because it's huge; it's three volumes; it's a total of maybe eight or ten inches of paper and printed two sides.
>
> Bob Reynolds, the air pollution control officer in Lake County,... he said, "Well, at least you could put the BTU rating on the cover page."...
>
> I said, "What do you mean?"
>
> He said, "When I throw it in the fireplace, I'd like to know how much heat I'll get out of it." (Krauss interview, KMD vol. 8, 366)

The reclamation of the McLaughlin Mine received national acclaim:

> **Homestake shows how good a mine can be**
> Seth Zuckerman
> Jan. 19, 1998
> McLaughlin Mine in Northern California.
> [T]he mine improved the land around it. The mining company has cleaned up three abandoned mercury mines, removed cows from most of the 11,000-acre site, and collected two decades of baseline scientific data.
>
> But most significant are the company's efforts at reclamation: Native oaks now grow atop waste-rock dumps and the mine has kept its heavy metals from worming their way into the downstream food chain.... Homestake has channeled about one of every 40 operating dollars into environmental efforts, amounting to some $2 million

a year when ore was still being extracted and waste rock piled for burial.

[W]hen the Mineral Policy Center prepared "report cards" on various mining firms in 1992,...Homestake walked away with the most positive one. The Mineral Policy Center's recent book, Golden Dreams, Poisoned Streams, says the McLaughlin Mine "demonstrates that hardrock mines can be both profitable and environmentally responsible." (Zuckerman 1998, n.p.)

While the staff at the mine site were busy with the shutdown and reclamation, the corporate structure in San Francisco changed. The company had acquired a Canadian firm, International Corona Corporation, and some Canadians served in new executive positions. At a meeting with the mine staff near the end of the shutdown, one of them said, "We've got to change the McLaughlin culture" (Krauss interview, KMD vol. 8, 396). This dismayed the employees who were proud of the spirit of cooperation across departments, the exemplary production levels, and nationally recognized achievement in safety and environmental protection.

Krauss was puzzled by a decision made in San Francisco,

that we should close down the administration building and move the remaining administrative and management personnel to the old truck shop offices.... [T]he truck shop...had been heavily used for fifteen years by people walking in the mud all winter and was full of dust and rattled by blasting, and it was in terrible shape, whereas the admin building was removed from those activities, in perfect shape, and all of those people were settled in, the computers were running, the accountants were working, and the bills were getting paid....

[I]t was a [five or six miles] shorter commute. It was tied into the communications system. In order to make the move, we had to expend a huge amount of money to extend phone lines and refurbish the building. I think the original estimate was that it would cost a few tens of thousands of dollars to make the transition, and it ultimately cost closer to $800,000 to close the admin building and move those people to the truck shop. (Krauss interview, KMD vol. 8, 395–396)

The Human Resources managers, Jay Wilkes and his successor Don Field, encouraged a positive attitude during the shutdown. Rather than bemoaning the fact that, “Oh, my God, I’m out of a job. Woe unto me,” it was, “I’ve worked for McLaughlin for fifteen years, and I’m the best there is in this business, and I’m a highly employable, marketable, skilled person, and people ought to want to hire me” (Krauss interview, KMD vol. 8, 394).

In the end, Purtell concluded that people believed that they were treated fairly; he was pleased that within a year after the layoffs, everyone was either employed or in training for a new job, and no suit was ever filed over layoffs. It was a remarkable record for a shutdown.

Enderlin looked forward to the next stage:

> And that’s the fun part, to see how we can transfer the McLaughlin Mine into the Donald and Sylvia McLaughlin Reserve. (D. Enderlin interview, KMD vol. 8, 87)

CHAPTER 20

THE McLAUGHLIN NATURAL RESERVE

One of the best-studied pieces of ground in the world

—Raymond Krauss, environmental manager

From the beginning, the plans to mine the Manhattan deposit for gold envisioned a facility for research and education when operations ended. Raymond Krauss, charged with obtaining permits, recalled that this was a selling point with James Hickey, a key local planner. It satisfied William Humphrey, Homestake vice president, who wanted to avoid additional expense or liability for the company. Krauss traced the formation of the plans:

> We developed a consensus in that committee [the Environmental Data Advisory Committee] about the post-mining land use of the site as an environmental research station, which was an outcome... that I was very pleased with.... I've always had a personal view that the academic world ought to play a more relevant role in resource management activities, and it seemed to me it was an opportunity to get the academic community involved with the project in a positive way.... So it had an underlying political benefit to Homestake as well as a land use that was viewed as beneficial by each of the three counties. (Krauss interview, KMD vol. 8, 279)

Homestake had connections with several local universities. When Jim Anderson directed the exploration drilling at the Manhattan site, he recruited geology students from Chico State University to collect and log the drill core. Ray Krauss, when he worked in the Sonoma County planning department, taught part time at Sonoma State University; he taught one class on environmental issues around geothermal development and mining and agriculture and forestry, and another on environmental impact analysis.

The connection with UC Davis developed slowly and was the one that survived. Before he joined Homestake, Krauss had served on the State Mining and Geology Board along with Robert Matthews, who was the first environmental geologist hired in the Department of Geology at UC Davis and who became associate dean of environmental studies in the College of Agricultural and Environmental Sciences. Matthews followed with interest the Manhattan project, regularly visited the site, and took students there for field trips. This also led to a connection with limnologist Charles Goldman in the same group at UC Davis, who conducted environmental monitoring on aquatic systems for the project in Yolo County.

The university welcomed the acquisition of the site, as Krauss recalled:

> The university was really very excited about access to the site because of its accumulated environmental data, starting with the development of the baseline data and the topographic mapping and geologic mapping and vegetation mapping and all of the surveys of rare plants and the twenty years or eighteen years of meteorological data and seventeen years of aquatic ecology data and water quality data.... [I]f you're coming in to study some particular aspect of ecology, having all of that information...already in hand saves you the trouble of doing it.... It's probably one of the best-studied...and characterized...pieces of ground in the world. (Krauss interview, KMD vol. 8, 409)

SERPENTINE. A ROCK OR MINERAL, OF A DULL GREEN COLOUR AND WITH MARKINGS RESEMBLING THOSE OF A SERPENT'S SKIN

California is the only state that has an official rock: serpentine. The website for the University of California Natural Reserve System (NRS) describes the McLaughlin Natural Reserve:

> Located at a former gold mine, McLaughlin Natural Reserve protects unusual serpentine habitats. Oak woodlands are interspersed with serpentine and non-serpentine chaparral in a mosaic that includes pristine habitats, rangelands, and reclaimed mining areas.
>
> The Homestake Mining Company has collected baseline data on the site's geology, soils, hydrology, air and water quality, archaeology,

and terrestrial and aquatic ecology; ongoing environmental monitoring adds to the computer database. Establishment of the reserve as a field station dedicated to environmental research is part of the long-term reclamation plan for the gold mine.

The site is visited by university courses in plant ecology, California floristics, geology, and creative writing.

Cathy Koehler is Director and Paul Aigner, Stewardship Director

Size: 2,776 hectares (6,940 acres)

Website 10/29/2020 lists five pages of scholarly publications.

In 2002 Professor Susan Harrison was the director of the NRS at the UC Davis, which included the Donald and Sylvia McLaughlin Natural Reserve, the former McLaughlin Mine site. She was born in Sonoma, California, in 1961; her father, a physician, was an environmentalist who fought a proposed freeway in the Sonoma Valley. As an undergraduate at UC Davis, she was the student member of the Campus Advisory Committee to the NRS. She earned a bachelor of science in zoology and a master of science in ecology at UC Davis, and a doctorate in biology at Stanford University. In 1991 she became professor of environmental science and policy at UC Davis. She lists her areas of expertise as ecology and conservation biology, especially of plants on serpentine soils, and ecology of California oak woodland, grassland, and chaparral. She recalls,

> In October 1984, as early as that, people on campus were thinking… maybe we should have a serpentine reserve. NRS didn't yet have a site that represents this very important habitat in California, with all the interesting endemic plant species and so on of serpentine. So various people, including me, including Phil Ward, now my husband, and several others formed a Serpentine Reserve Subcommittee of the campus NRS Advisory Committee. (Harrison interview, KMD vol. 8, 120)

In February the next year, Ray Krauss spoke to the NRS Advisory Committee, telling them how environmentally sound the mine was, and showing its potential as a reserve. In June 1985, the committee visited the site. They visited five others before deciding that the one they called Morgan Valley was the best, as Susan Harrison recalled:

> It was the only one that had full range of habitats, all nine of the butterfly species, just generally really high diversity of the flora and

> fauna. That's the site where Morgan Valley Road crosses the Napa-Lake County line. It's the place that we now call Research Hill.... There's a serpentine wetland there, there's a hill that has a whole series of springs that feed little streams that come down, there are a lot of rare plants that are concentrated in this one place,... [a]nd there's a big grassland across the road. (Harrison interview, KMD vol. 8, 122)

IT'S FUN TO BE DOING SOMETHING SO BASIC

In 1997 Susan Harrison became the faculty coordinator for the entire NRS, chairing the advisory committee and reporting to Kevin Smith, the vice chancellor for research. The Sixteen Fire of October 16, 1999, burned about 40,000 acres, including the McLaughlin Reserve. Fortunately, the buildings were spared. It was a research opportunity for Harrison:

> Even though people have written a lot about the floristics of California serpentine plants...nobody had really done anything on fire ecology and serpentine.... That was surprising, because in non-serpentine chaparral it's well known that there are many species that you only see after fires, so-called fire followers whose germination is stimulated by fire in some way.... [T]here's a lot of species that just can't grow there until a fire comes along and opens things up, and then these herbs, wildflowers come in and flower for a few years, put a lot of seeds into the soil, and then the chaparral closes in again and you don't see those species again for another fifty years, or until the next fire. (Harrison interview, KMD vol. 8, 154)

After the fire, Harrison and postdoctoral student Hugh Safford studied its effects:

> In the spring of 2000 we went out and established plots in burned and unburned serpentine and non-serpentine chaparral and grassland. I studied 100 grassland sites, and Hugh studied 80 chaparral sites. We're basically asking the question, How much does fire affect the plant diversity within chaparral and grassland? How do the effects of fire differ between serpentine and non-serpentine habitats?
>
> We're hoping this will be a definitive paper on the subject, since there haven't been any others. It's fun to be doing something so

> basic. It isn't often in ecology these days you get to study some really basic thing like that, because a lot of that kind of natural history type work has already been done. But in this case, it was just...a hole in our knowledge. (Harrison interview, KMD vol. 8, 155)

In December 2000 Susan Harrison spoke of the widening research at the McLaughlin Natural Reserve, saying it was "no longer just bugs and bunnies" (Harrison interview, KMD vol. 8, 168).

> [At] Davis we've had...this program called Nature and Culture, which is for students that are looking at nature from both the humanities and sciences perspectives.... The...class goes out there for a week and really studies the area.... [A]ltogether maybe a dozen different classes from Davis have used it, either for a day or for several days.
>
> And classes from San Francisco State and Berkeley and a few other places have gone there. Occasionally we get a geology class there.... [A]t this point we have several dozen people doing research there, which is great.... The place has started to pick up a lot of momentum. More and more people on campus are aware of it, so they get attracted to it as a place to work. (Harrison interview, KMD vol. 8, 168, 169)

MY UNDERLYING OPINIONS BEGAN TO CHANGE

Susan Harrison's ideas changed from her first visits to the McLaughlin site, when she says she "fell in love with the place" (Harrison interview, KMD vol. 8, 122) with its serpentine and many butterflies:

> I was very unhappy to hear that a mine was going to come to this area.... After I left the area, went elsewhere for a few years and then came back to Davis as a faculty member,...I avoided going out there for a few years because I remembered what it was like before the mine was really visible, and I didn't really want to go back out there again and see it all changed.... [W]hen I did go back out and saw... the big buildings...I wasn't exactly thrilled.... [The] whole area... is a wonderful place to do research. I was still not completely converted to the cause of the UC and Homestake being partners. (Harrison interview, KMD vol. 8, 172)

When she became involved in managing the reserve system, her ideas began to change:

> [I]t still took a while before being out there and interacting with people before my real underlying opinions began to change.... I've been really impressed by the people at Homestake that I've interacted with—not just Ray [Krauss] but Dean [Enderlin] and Scott Moore, the surveyor, and Sarah Robertson...Pat Purtell,...and others whom I've met out there.... Just the sense that everyone there is really dedicated to their job, and there's this sense of professionalism and everything being on the up and up. (Harrison interview, KMD vol. 8, 172–73)

Homestake had a good reputation that helped to convert her:

> [Over] the years...I've talked to an awful lot of people that have come into contact with Homestake—whether it's the employees or the neighbors or the government regulators or the environmental groups that have been keeping an eye on what's going on out there—and I've virtually never heard anybody say anything bad about Homestake, and that seems pretty remarkable. (Harrison interview, KMD vol. 8, 173)

IT'S SOMETHING WE'RE ALL RESPONSIBLE FOR

In the end, she agreed with Ray Krauss:

> Of course, any rational person has to look at a mine and feel like they are partly responsible for it. If they're a user of metals, then it's something we're all responsible for, not just the company that built the mine. So on that level accept the idea that the mine is there and also it's a good thing that they have tried to set a standard for how to do mines right, so in the end it has taken a while, but now I'm pretty much of a convert to the point of view that Ray has been doggedly advocating all along: that mining is important and it's important to do it in a way that sets a better standard for mines in general. (Harrison interview, KMD vol. 8, 173)

In January 1993 a signed agreement between Homestake and the regents of the University of California created the Donald and Sylvia McLaughlin Reserve. In 1998 a modest overnight facility was set up in an unused warehouse and was named the Ray Krauss Field Station. At the same time, when Homestake allowed the NRS to begin managing the natural areas of the property as a reserve, use of the reserve for studies of ecology, evolution, and other environmental sciences grew rapidly. For four years, an active mine coexisted with an increasingly active environmental research station.

In 2001 Sylvia McLaughlin wrote to the author:

> Don was very enthusiastic about the plans for developing a mine and, of course, he was so pleased that it was named for him.... With Don's interest in education and in the University of California and with my interest in the environment, creation of the Donald and Sylvia McLaughlin Reserve is a wonderful new use for this extensive mining property. It is so good to know that the Homestake Mining Company, by an agreement with the University of California involving a large acreage of the McLaughlin Mine, will provide an ideal area for students of the Davis campus to do many kinds of scientific research. This special, natural reserve will also provide inspiration for students to do creative work in the humanities. I hope this educational use of a former gold mine property can be a model for other mining and industrial sites (pers. comm.).

On April 24–25, 2007, when she was ninety years old, Sylvia McLaughlin and Eleanor Swent spent the night at the nature reserve's Raymond Krauss Field Station. Susan Harrison led them on a tour of the facilities and the terrain and Sylvia expressed her satisfaction and pleasure. McLaughlin East Shore State Park, extending eight and a half miles on San Francisco Bay from Oakland to Richmond, was named after her in 2012.

Sylvia Cranmer McLaughlin died on January 19, 2016, at her home in Berkeley, California, at the age of ninety-nine. Berkeley Mayor Tom Bates said in a eulogy, "Words are hardly adequate to convey [McLaughlin's] profound influence on protecting the environment" (*East Bay Express*, January 20, 2016).

In May 2018 Professor Susan Harrison was elected a member of the National Academy of Sciences in the section on Environmental Sciences and Ecology. The announcement in the UC Davis *University News* reads:

Susan Harrison, a professor in the Department of Environmental Science and Policy, has been elected to the National Academy of Sciences, one of the highest honors a scientist can receive. As an ecologist, Harrison studies the processes that shape and maintain plant species diversity at the landscape scale....

In a 15-year study at UC Davis' McLaughlin Natural Reserve (between Lake Berryessa and Clear Lake), she noted how a loss of plant species richness, especially of native wildflowers, is tied to drier winters such as those experienced during drought. She reported on her work in a paper titled "Climate-Driven Diversity Loss in a Grassland Community," published in the Proceedings of the National Academy of Sciences in 2015. (Stumbos 2018, n.p.)

CHAPTER 21

LAKE COUNTY, LATER

Homestake was a breath of fresh air.

—William Cornelison, Lake County superintendent of schools

William Cornelison, born and educated in Southern California, was high school principal and school superintendent in two Lake County districts, Middletown and Konocti (which includes Lower Lake), from 1979 to 1994, and then was elected as Lake County superintendent of schools. He saw the impact of the McLaughlin Mine on the schools and surrounding communities. In 1979 the Middletown school district had about three hundred students and served a community of about fifteen hundred people. By 1986, when the mine was in full operation, the school population had quadrupled. In 1995 he recalled,

> Many of the people that came out to Homestake were college graduates. They were engineers, people who had high desires for their students, as compared to the more settled agricultural population that we were experiencing.... it inspired lot of other students.... That was a very positive influence.... It will certainly be missed.... [I]t was no doubt that Homestake was a breath of fresh air." (Cornelison interview, KMD vol. 2, 277, 278, 279, 290)

He was grateful for help from Robert Reveles, Homestake's vice president for governmental affairs, in setting up a lasting program for parental involvement in the schools:

> We developed...a weekly folder that was sent home for the children in grades K through 8.... [It] was a very brightly colored folder that had all the student's work in it for the week, plus information for the parents. We got parents to look for that every Tuesday, as a practice. It's still being continued over in the Konocti district. And the Home-

stake logo is on that folder, indicating that it is a program supported by Homestake Mining for the people.... [I]t did make a difference. (Cornelison interview, KMD vol. 2, 288)

Homestake funded the training for construction workers, and the industrial-tech program building at Yuba Community College in Clearlake is a lasting reminder of that contribution.

Cornelison anticipated the impact of the mine's closing:

> [After the mine closes,]...there will be no severe impact. It's not going to just all of a sudden shut down one day and not be there anymore.... We're a highly transient county. People come and go here an awful lot,...so we're used to changing enrollments and students coming and going. This is typical of a low-income population. People are looking for work, and if you live in a mobile home and you rent, you're highly transient, and you'll be here today. You can pack up, be gone tomorrow, literally.... [I]t's an Anglo migrant population, not associated with farm workers at all. Just people looking for any kind of work and will go wherever they can go. So we're used to that. (Cornelison interview, KMD vol. 2, 290)

Ray Krauss recalled the celebration when the first million ounces of gold had been produced. He organized a celebration, part of maintaining public support for the operation. Everyone from the three counties who might have had a hand in the project was invited. There was a barbecue at the picnic area, and everyone had a chance to have their picture taken holding a gold bar. Jack Thompson had a medallion struck, gold plated and encased in plastic, that was given to the guests as a memento, and Krauss recalled the pride of the employees at showing their families where they worked. He realized that in Lake County that pride lasted through the life of the mine and beyond.

And Knoxville was preserved:

> [T]he Knoxville Ranch is now the Knoxville Wildlife Area, owned and managed by the state Department of Fish and Game as a significant unit of the Blue Ridge-Berryessa Natural Area. (Krauss interview, KMD vol. 8, 418)

Claire Fuller, owner of Fuller's Superette on Morgan Valley Road, had a booming business during the construction period and then saw the decline as the workers became settled. In 1995 she recalled some lasting benefits from the mine: the new streetlights and two traffic lights in Lower Lake:

> and then, like I say, paving the street, and they built a whole new road out to the mine. (Fuller interview, KMD vol. 3, 130)

She recalled earlier times when there was a branch bank in Lower Lake:

> It was a Bank of America branch. It was either '85 or '86 that they closed.... There was a lot of people very upset.... [P]eople came into town for the post office, which was a block from the bank, and they went to the bank. Then they came to the store, did their shopping. Everything was right in that three-block area there.... You would go uptown in those years, and you couldn't find a place to park. In fact, now it looks like you can't find a place to park sometimes, and I don't know what they're up there for, but there must be something there still.... [I]t's a nice little town. (Fuller interview, KMD vol. 3, 133, 134)

Beverly Magoon, another merchant in Lower Lake, echoed regret for the bank:

> Well, the mine put in a road. That made a lot of difference.... When the bank closed,...that was really devastating to everyone around here. (Magoon interview, KMD vol. 5, 196, 216)

James Jonas, fuel distributor, was another one whose business was affected by the mine. In 1997 he reflected on what the mine's closing would mean:

> Homestake is running its course, dropping down.... [Y]ou've got some other things replacing it like vineyard development, but that's a drop in the bucket compared to what we had there. So property values here in the county have dropped considerably.... Lake County is still growing, and there's people coming in but it's more of a retirement base.... [T]he Homestake thing really did jack up the economy tremendously and it's gone away.
>
> ...And we've gotten an infrastructure here that we probably would have never had if it weren't for Homestake and the Geysers. You've had the Yuba College Junior College program.... Well, they

still have...the facilities there, and they're training...to do other things now.... [O]ur economy will bounce back; it's doing that right now. (Jonas interview, KMD vol. 4, 211)

He was adjusting his business to the change:

Well, it's been an interesting ride, I'll tell you that.... [W]e're carving out a little niche with...heating and air conditioning contractors...that will enable us to stick around...and maybe not in the fuel business.

[T]hat corner down there is probably one of the highest traffic corners in the county.... [W]hen I go out to the stop sign and have to wait for five minutes for all this traffic going by I say, "Where did all these people come from?" (Jonas interview, KMD vol. 4, 216, 217, 218)

He praised a local volunteer effort:

[T]here is the old historical schoolhouse museum.... My aunt, my mother's sister, is part of the Historical Society, and I called them the Hysterical Society, but it didn't go over too well.... That place was so deteriorated.... [A] few of the local people...got in there and...they've done a marvelous thing with that place.... My wife, and I, and a couple of friends went to...The "Son of the Sheik,"... the silent film with Rudolph Valentino up there...and had a little black tie dinner down at the Brick Hall across the street beforehand, and period costumes, and that kind of thing.... The work that...the volunteers have done up there is just outstanding.... Now we go... to the theater up there and have a good time.

...Anyway, to make a long story short, the mines added a tremendous amount to this community, and some of it is staying with us; its passing has caused an economic downturn to a certain extent, but that's something that's survivable. I think their presence here has given more than they ever took. (Jonas interview, KMD vol. 4, 219, 220)

Walter Wilcox, Lake County Supervisor, assessed in 1995 the impact of the mine:

We were grateful that they came, we're grateful they're here. Yes, it is going to have an impact on us when they leave, because it is employment, and those people spend money here in all of our businesses,

> and those businesses pay taxes, those employees have bought homes and will be probably moving on, leaving that void. And all of their houses will have to be taken over by the banks if they're unable to fulfill their obligations. You see more and more of that in the paper every day. (Wilcox interview, KMD vol. 7, 395–96)

Bonny Hanchett, editor of the *Clear Lake Observer*, recalled economic benefits:

> There was no lodging at the mine. So they found places to live, and that helped the resorts; and, of course, they had their payroll, and that was quite a substantial payroll because all of the processing went on in Lake County, so most of the people that worked in that end lived in Clearlake or Lower Lake. So it had a very positive impact. (Hanchett interview, KMD vol. 4, 86)

She paid tribute to Bill Wilder:

> Another major impact was that Mr. Wilder, who owned the mine, who benefitted financially, has been quite a benefactor for Clearlake. He built the cinema, and he gave the town the property for the new library.... [W]ithout him, I don't think we would have much of a library today. So that was an offshoot of the mine that had a really good impact on the community as a whole. (Hanchett interview, KMD vol. 4, 86)

William Cornelison also expressed gratitude for Bill Wilder:

> [W]e've gone to Bill with ideas and thoughts,...and he has responded so beautifully to the community needs in building the theater, serving the needs of young people.
>
> The theater is the major attraction in Clearlake.... Any time that we need to use the theater, for example, when "Schindler's List" came out, the producers...sent the instructional material and encouraged all the high school students to go see that movie so that they could have a better understanding of the historical facts surrounding the Holocaust and the impacts. It's a very powerful film. So we went to Bill and said, "Would you be able to do this?"
>
> "Sure, no problem, no problem." I think he charged a dollar a stu-

dent, just to offset the cost of opening the theater up during the day and hiring people to come in, run the projector and do all of that. But, yes, we sent the whole population of Lower Lake High...over there, and I think the eighth graders from Oak Hill went over.

So any time you ask Bill for anything, he's been very supportive. I sense he feels that Lake County and the community has been good to him. He, in turn, has been good...to the children, to the people in the county. (Cornelison interview, KMD vol. 2, 297, 298)

Cornelison recalled another example of Wilder's generosity:

[T]hey had a small, little library over at Clearlake. It was no bigger than a couple of portable classrooms.... And they...had a chance to get a library grant. The population is about three times the population at the south end of the county than it is over here. And Lakeport has a big beautiful library. They could get the money to build the building and all that, but it cost them a lot to buy the land. So they went to Bill and said, "We've got this problem."

And he says, "If the town needs a library, there's the land." And basically it was no cost; he just donated it and helped them out so that the funding from the grant was sufficient. And now they have a beautiful library. It's as big as the one here at Lakeport. (Cornelison interview, KMD vol. 2, 299)

In 1995 they were asking for Wilder's help in another project:

We're...dealing with him right now in getting some land to set up our court school program.... The county offices run classes for students that get expelled from the schools, so they have a place to continue their education. That's a more controlled setting than their regular school...the students are under probation.... It's a very successful program. It rehabilitates a lot of kids. It gets them straightened out...Bill is working with us now to find some land. (Cornelison interview, KMD vol. 2, 298)

John Drummond, Lake County Schools attorney, also evaluated the impact as the mine operation wound down:

Now of course we are seeing what we learned in economics class: for each industrial worker there are five service jobs created. We are feeling that effect; it is really noticeable that as the mine winds down

> and the mine workers leave, the service workers are beginning to go too. (Drummond interview, KMD vol. 2, 289)

Dennis Goldstein, Homestake attorney, recalled dealing with Wilder in the final negotiations to buy the One Shot Mine property:

> It was always a pleasure to deal with Bill. He's a gentleman and just a wonderful guy.... [T]here are very few people on earth who could... become as wealthy as he did and still stay the same genuine, sincere, warm, easy-going person.... He's just a wonderful fellow.... He loved the property. He loved the ground.... We bought him out... [and there] really wasn't any problem with that. (Goldstein interview, KMD vol. 3, 207)

Wilder had his own recollection of the sale:

> I think there was a lot of people on the board of directors that were hoping I'd have some kind of a fall or something,...because I was fairly hard to deal with I guess...I don't see that I should be putting my whole life into something and then giving the whole thing away. I said, as long as there's enough to sustain both of us, I think that my people are entitled to something, too, because they did just as much on the development of this thing.
>
> ...Well, I projected it and I want about $5 million for half of it. I'm figuring the total thing is going to be about $10 million.... Oh, jeez, they thought I was loony. But then they finally decided it was worthwhile. History proves this to be a very smart move. (Wilder 1996, 68, 69)

Bill Wilder moved from the site of the Manhattan Mine, bought property in Upper Lake, and became an admired and respected member of the community.

Della Underwood praised his lack of pretense even after he became wealthy:

> I'm sure he didn't have any idea there was all that money sitting under there.... He's the same old Bill, though. (Underwood interview, KMD vol. 7, 335)

Claire Fuller felt the same about Wilder's daughters:

> His daughters used to shop in the store with us. You'd never think they were daughters of a millionaire.... They're nice kids.... They were just ordinary people. (Fuller interview, KMD vol. 3, 136)

She also praised Bill Wilder:

> [H]e has given a lot to the town of Clearlake, you know. And he's always right there when they need something.... Put up a nice building, a nice theater. (Fuller interview, KMD vol. 3, 136)

IF YOU'RE WORKING MERCURY, IT'S NOT ANY DIFFERENT THAN IF YOU WERE A LION TRAINER

Bill Wilder's final interview was at Walter Wilcox's Redbud Motel in Nice, California, on the shore of Clear Lake, on June 23, 1995. It began as a wide-ranging recollection of the history of the One Shot Mining Company and other mines and miners in the district.

He scoffed at the exaggerated talk of the dangers of mercury:

> I think it's all baloney.... I mean, jeez, all these years, and I don't know anybody that kicked the bucket from mercury, see?...I've got my original teeth and I'm seventy-one, so that's not so darn bad. See, when you pull all these into perspective, you kind of go, Wait a minute. Those are generalities that aren't true. (Wilder 1996, 127, 128)

He was always careful:

> It can happen. If we would get high in mercury, my teeth would bother me. I'd ache a little, you know, at One Shot, and...I'd get a blood serum thing, and sure enough, I'm getting high, because I'd be getting reckless. And that's normal, you...kind of get contempt for something until it bites you a little, and then you go...back off from it. I'd...start watching myself,...I'd wash my hands.... It's just little things. It isn't one big time that you do it; it's over a steady period of time your hygiene degrades, because you say, "Oh, nothing's happening to me." Well,...it didn't hurt my teeth any, it just gave me a warning. (Wilder 1996, 128)

He was realistic about the risks:

> [I]f you're working mercury, it's not any different than if you were a lion trainer and you said, "Ah, don't worry about them lions," and

> you go walking into the cage and this thing takes your head off, you'd say, "You didn't use your head. The lion took it off."...You have to be cautious.... "Holy smokes, this can hurt me."...You knew that, and if you don't, you shouldn't be fooling with it.... That's where you say, "This is my body, I'm going to take care of it." (Wilder 1996, 128–29)

I DON'T WANT ANY THANKS

Then he talked modestly about his many donations, the Clearlake firehouse among them:

> Up in Clearlake, I have another building, have the firehouse there. See, when they built the new firehouse, a guy bought the old firehouse, but he didn't have the dough for it, and they wanted to get money to build the new one, so they talked to me and I bought this guy's note, and then gave them $100,000 so they could build the new building. They built it, they're an all-volunteer company; a nice firehouse over there. (Wilder 1996, 121)

He also helped the police department:

> I said, "I don't want any thanks from this, because...I've got property there, and the police department is vital for everybody."... That's why I helped the police department there, because they've got a tough row to hoe. It's not easy at all. (Wilder 1996, 123–24)

He was gratified that Homestake had done well, and men who were managers at his mine had gone on to be top executives:

> This thing here is, I think, a pretty good mine, pretty good and big, and then I just look at where we started out here: you got Jack Thompson, he's president of the company. Where'd he come from? Here.... And Ron Parker.... [T]his hasn't created a bunch of dumdums.... And that's kind of neat. And it's been great for me, for a guy with my education and background, and I can say, "Hello, Jack, how are you?" and...he'd say, "Hi, Bill, how are you doing?"...And Ron Parker, we're good friends.
>
> And I don't think that the company has done bad. I'd feel bad if I sold them a pig in a poke, and I don't think I did. (Wilder 1996, 132)

I THINK ODDS ARE ON MY SIDE

He still hoped to find another mine:

> [A]bout a month...ago one Sunday...[I] took some sandwiches and just went out to Bartlett Springs and—you know, I got some theories on that; there's an intersection through there; there's a strike. But by golly I picked up some...chalcedony out there and nobody, I don't think, has ever really looked at it. And I know that if my theory is right, if I can pick up chalcedony about where I think it's going to be, within a half a mile, I'm not too darn far off anyhow....
>
> ...Every place there's a gold mine, there's not one gold mine, there's a bunch of them. So we've got a place that's only got one. Well, it's the only one in the world, then, that has one gold mine. I don't believe it. I mean, I think odds are on my side.... In other words, it would be an aberration, and I don't think that's true. You're right, I'm an optimist. (Wilder 1996, 119, 126)

He summed up his philosophy:

> Well...I think I did a good thing with my money.
>
> That's a good thing.... [T]hat's philosophical, but...almost all people want to have something, that you did something while you were alive. If you're just living here like a redwood tree, you know? Unless you can live to be 2,000 years old and be the biggest redwood tree in the world, then you're famous, but if you're any other redwood tree, they cut you up and make two-by-fours out of you. (Wilder 1996, 123, 133)

Everyone who knew Bill Wilder agreed that he was not just living like a redwood tree, that he did many good things with his money, and really did something while he was alive. His One Shot at the Manhattan mercury mine had hit the bull's eye.

> Lake County Record-Bee obituary March 1, 2006
> James W. (Bill) Wilder, age 81, passed away Feb. 23, 2006, in St. Helena. He was a 40-year Lake County resident. Bill was born April 4, 1924, in San Francisco. Bill was a miner and a contractor for over 60 years.
>
> He served in the Army Air Corps during World War II. Bill was a member of the Lions Club, SAE (Society of Automotive Engineers), AIME (American Institute of Mining Engineers), SME (Society of Mining Engineers), American Legion in Clearlake and the VFW (Veterans of Foreign Wars).

...He is survived by...[four children,] eight grandchildren and four great-grandchildren. (*Lake County Record-Bee*, March 1, 2006)

...The funeral service is...at Clear Lake Memorial Chapel in Lower Lake. A graveside service will be at the Lower Lake Cemetery....

...Donations may be made to the Salvation Army or an organization of your choice. ("James W. (Bill) Wilder" 2006)

In 2001 the San Francisco office of Homestake Mining Company closed.

The final shareholders meeting was held on December 14, 2001, at the Walnut Creek Marriott.

Barrick Completes Merger with Homestake
TORONTO—(BUSINESS WIRE)—Dec. 14, 2001—Barrick Gold Corporation (NYSE:ABX TSE:ABX. PARIS:ABX.) (SWISS:ABX.) (LSE:ABX.) today announced the completion of its merger with Homestake Mining Company (NYSE: HM), strengthening Barrick's leadership position as the most valuable gold mining company by market capitalization. "Going forward, our goal is to be the most profitable, lowest-cost producer, not to be the biggest producer," said Randall Oliphant, Barrick's President and Chief Executive Officer.

Homestake shareholders strongly approved the merger during a special meeting held earlier today. Under the terms of the merger agreement, each share of Homestake common stock was converted into 0.53 Barrick common shares. The combined Company has approximately 536 million shares outstanding. Former holders of Homestake common stock now own approximately 26% of the outstanding Barrick common shares.

"This merger ties strength to strength," said Mr. Oliphant, "creating a company with a stronger balance sheet, stronger free cash flows and more opportunities than either company had alone." Jack Thompson, Homestake's Chairman and C.E.O., said: "We are pleased that our shareholders have shown overwhelming support for the merger, which uniquely positions the combined Company for an exciting and dynamic future." (Borg 2001, press release)

POSTSCRIPT

The Homestake gold mine in Lead, South Dakota, is now the Sanford Underground Research Facility, administered by the South Dakota Science and Technology Authority. This builds on the legacy of Raymond Davis Jr., a Nobel laureate for physics in 2002 for his successful detection of cosmic neutrinos in a tank holding 100,000 gallons of perchloroethylene that was installed in 1966 nearly 5,000 feet underground, at the Homestake Mine. Research continues there today, in projects such as the Compact Accelerator System for Performing Astrophysical Research and Large Underground Xenon dark matter detector.

More than a few of the Homestake Mine employees were second- or third-generation workers for the company and mourned its demise. Even more recent employees appreciated its long history. Mark Stromberg worked first as a consultant on closure planning for the Homestake Mine, and then, in early 2001, he was hired as manager of the Homestake Mine Reclamation Project. Mining ceased later that year. On January 18, 2002, the last gold ore was processed, and on February 13, 2002, the mill was shut down forever. Stromberg sent this email in 2004 with the subject line "Sense of History":

> I went to a small party on Friday night, and that evening I heard a sad commentary on culture and companies.
>
> One of the guys worked for Homestake at various jobs for 23 years and most recently had operated the Yates Hoist. The hoist controlled the cables on which the man-cage hung and which was used to go up and down the 4,850-foot mine shaft. As you drop down the shaft the side of the cage is open and you watch the rock walls whiz by. Greg operated the hoist on the last June day it was used and brought the cage up for the last time. He told me that on that historic day, the last cage after 126 years of mine operations, there were only two people to bear witness. Greg pulled the levers and watched the

gauge as the arrow showing the elevation of the cage monitored its progress to the surface. An avid amateur photographer took a few pictures. And nobody from Barrick management bothered to be there! Greg shut down the hoist (one of eight in the world), cleaned out his locker, shook his head, and went home. He is now driving a milk delivery truck, leaving Rapid City at 4 AM.

Where is their sense of history? Where is their understanding of workers and their connection to the mine and to work? (Personal message to Eleanor Swent, 2004; quoted with Stromberg's permission)

APPENDIX

Homestake Mining Company McLaughlin Mine Tour Guide Script, Version 1993–01

Compiled by D. Enderlin, Geology Dept.
(Original Windows Word document: HMCTOUR.DOC)

Introductions: Tour Guide and Bus Driver

Good morning (afternoon) and welcome to Homestake Mining Company's McLaughlin gold mine. We appreciate your interest in the project and want to thank you for purchasing your tickets through the Lakeport Chamber of Commerce. These tickets are tax-deductible, charitable donations, and any proceeds over our expenses will go into a fund for various area programs and nonprofit organizations. Cameras are welcome on the tour, and please feel free to take pictures wherever you wish. No food and beverage consumption or smoking is allowed on the bus. Also, please make use of the garbage can up front. We are very ecologically minded here, and trash, either on the bus or on the grounds, is frowned upon. Thank you for your help!

Our first stop will be the lunchroom in back of our administration building.

This is the Administration building. It contains the office of the resident general manager and various administrative departments, including accounting and personnel. Just beyond the administration building is the maintenance building and warehouse. As with mines both old and new, the shop in the maintenance building is an important hub of activity. It is well equipped to fabricate or modify many of the hard-to-find parts necessary to keep our complex machinery in operation.

In modern gold mining operations, such as the McLaughlin Mine, huge tonnages of rock must be processed daily in order to recover enough gold and silver to make a profit. This is done in three major steps:

1. Mining, where gold-bearing rock is removed from the earth.
2. Processing, where the rock is pulverized and made ready for extraction of the gold.
3. Recovery, where the gold is dissolved from the powdered rock and ultimately poured into bars.

Each of the three steps requires careful planning, skilled operation, and adherence to strict environmental impact guidelines. During the course of today's tour, we hope to show you how these steps all come together to make the McLaughlin Mine a safe, efficient, profitable and environmentally conscious world-class gold mining operation. If you have any questions or miss something, please feel free to ask at any time.

Homestake Mining Company video presentation

[The introductory video contains sufficient information to acquaint people on the tour with Homestake Mining Company, its size, and its history.]

> In 1989, Homestake Mining Company produced 1,014,702 ounces of gold. The McLaughlin mine contributed nearly 284,000 ounces toward this annual total.

[Return to bus] [Drive to Model Room]

Model Room

These are the models of our crushing, milling, and processing facilities. Their construction was subcontracted by the Davy-McKee Corporation, who did the engineering work here in 1982 and 1983. They were built in Danville at a cost of almost $300,000 and were shipped here in 1983. In 1988 the model underwent a major modification to include details for a new $25 million facilities expansion program. This program was completed early in 1989. With ongoing and recent modifications to the models, it is estimated that Homestake's investment in them now totals $450,000! All in all, the models have definitely paid for themselves. Architectural modeling is used extensively in industrial planning. This is because three-dimensional planning is easier to do. It allows the engineers to see better how things work and fit together. The scale used on these models is ⅜-inch equals 1 foot. They are also helpful in safety and accessibility planning so things can be built with allowable headroom and adequate spacing in mind. Besides showing how things fit together, the models are useful for giving tours; training operators; showing specific areas to vendors, consultants, and technicians; and for troubleshooting various technical problems. Every element of the model is color-coded. Finally, its biggest savings comes in the inventory area. As you can see, everything is numbered, so there is no need to dig around in the files or out in the plant looking for replacement names and numbers. Things are not set out in here as they are on the property. Instead, they are arranged so that people can move easily among them to study specific areas.

[Point out key areas in model that will be seen by the tour] [Return to bus]

Process Area Drive-Through

As we pass beyond the machine shop and model room building, we can see the large structure that houses the autoclaves. This is the heart of the facility where the gold-bearing slurry (we'll learn more about that later), which is pumped from the mine, is processed. Ore slurry entering the complex is thickened and then preheated, using by-product heat generated from the autoclaves with supplemental heat from a boiler. In this same step, known as preoxidation, sulfuric acid is also added. This mixture is then pumped into the autoclaves, where oxidation occurs.

[Stop near front of pressure oxidation (autoclave) building]

Oxidation is the breakdown of sulfide minerals, such as pyrite (fool's gold). The purpose of this step in the process of gold extraction is to liberate microscopic gold particles that remain encapsulated by these sulfide grains, even though the ore has been crushed to consistency of talcum powder. Oxidation decomposes these sulfides, and exposes the gold in preparation for the C.I.P. circuit. To oxidize the ore, the slurry acid mixture is pressurized (to 330 psi [per square inch]) and further heated (to 400 degrees F[arenheit]) in the presence of pure oxygen. In this highly corrosive environment, breakdown of the sulfides occurs in about 90 minutes. The process gives off heat, which is partly captured by heat exchangers and reused to preheat incoming slurry. Another by-product of the oxidation is sulfuric acid, which is also recycled. To accomplish the ore oxidation, we use huge vessels called autoclaves, which are basically large pressure cookers. We have three of these, which operate continuously year-round (with the exception of maintenance periods). Each autoclave is 14 feet in diameter and 52 feet in length. The outer shell is 1.5-inch-thick steel, lined on the inside with lead and two layers of fire brick. Each was forged in one piece in Germany, then shipped to the Netherlands for lead lining. From there, the autoclaves were shipped to the port at Sacramento, where they were placed on special transport vehicles for shipment to the McLaughlin Mine site. The trip from Sacramento took two weeks at an average speed of 8 mph [miles per hour]. They are the heaviest legal loads ever transported on California highways.

[Continue past front of pressure oxidation building,
and pause at entrance to oxygen plant]

Oxygen for the autoclaves is produced on site at our oxygen plant, which is the tall white silo-like building to the right of the autoclave complex. Here, oxygen and a number of other gases are extracted from the air using a cryogenic (cooling) process. This plant is capable of producing 300 tons of oxygen per day. Once oxidized, lime is added to the acidic slurry to neutralize it in preparation for the final step in the extraction process: Cyanidation. Sodium cyanide (NaCN) is a compound composed of the elements sodium, carbon, and nitrogen. When dissolved in water, it becomes a very effective solvent of gold and silver. That is, it will actually dissolve the metals out of the slurry. Before the introduction of the cyanidation process in the 1890s, gold and silver mines used mercury (quicksilver) to recover the metals from the crushed rock. This process was known as amalgamation. Amalgamation was a very inefficient process, and the volatile element, mercury, was difficult to contain safely. As a result, most modern gold mines use cyanide. Cyanide is surprisingly safe to work with, if handled in a responsible manner. Yes, it is the same substance that is used in the gas chambers, but gold mines don't handle it in the same manner. In a gas chamber, the cyanide is combined with a strong acid, which causes the deadly hydrogen cyanide gas (HCN) to form. Cyanide at gold mining operations is only used under safe highly alkaline (low acid) conditions. Under these conditions, the process allows us to safely and efficiently extract nearly 95% of the gold out of the oxidized rock. Sodium cyanide is transported to the mine during the dry months in specially constructed cannisters called flo-bins. These can be seen immediately ahead of us as we round the corner past the oxygen plant, heading toward C.I.P.

[Pause at C.I.P. circuit]

C.I.P. stands for Carbon In Pulp. It is here that the gold-bearing slurry and cyanide solution are mixed. Carbon (activated charcoal) chips are added along with the cyanide solution to the oxidized slurry (pulp), giving rise to the name of the process. The carbon we use is derived from coconut shells! The carbon is a very special part of the gold extraction process, because once the gold has been dissolved by the cyanide, it is very hard to extract it from the solution. Carbon possesses a special property that causes the solution to give up its metals, much like the charcoal filters we use for our drinking water. Because of this unique quality of carbon, the C.I.P. process actually involves two chemical reactions happening at once: the cyanide dissolves the gold, but as soon as it does, the carbon grabs the gold from the cyanide. When cyanidation is complete for a given batch of slurry, the carbon (now laced with gold and silver) is separated from the barren slurry by means of screens. There are actually two sets of C.I.P. tanks. One is devoted to slurry that has passed through the autoclaves. This is called the sulfide circuit. The other is devoted to slurry from ore that was naturally oxidized before it was even mined, and that doesn't require the autoclave process to increase gold recovery. This special circuit was constructed in 1988 during our $25 million plant expansion, and is called the oxide circuit. What happens to the cyanide? The small amounts that are not reused in the C.I.P. circuits are rapidly broken down into inert compounds by sunlight, and are safely discharged with the barren slurry. The wide valley to our right is known as Quarry Valley. It is here that the slurry which has had gold extracted from it is discharged. Notice the red color of the material filling it. This material is called tailings. Tailings are the residue left over from the processing of the ore. It's what is left after the metal is extracted, and consists of water and the powdered rock pulp. Also in the water are residual chemicals used in the process. Such components, although considered non-hazardous, could degrade the local water quality and the environment if dumped or stored indiscriminately. In order to provide adequate containment for the tailings, a suitable site had to be selected. The current tailings impoundment was chosen because of its ideal features. It is a broad valley, with a narrow, easily dammed canyon draining it. The underlying bedrock is serpentinite, a rock that is extremely impermeable and perfect for holding the fluids in tailings. Together, these features were ideal, and allowed us to create this zero discharge reservoir. In every way, it is the perfect site for the impoundment. When milling operations come to an end, the tailings area will be covered with topsoil and revegetated.

[Turn left and continue past refinery]

Gold-bearing carbon chips from both circuits are collected from the screens, and treated with a concentrated solution of hot caustic soda (NaOH) and sodium cyanide. This chemical combination strips the gold and silver from the carbon, and once again dissolves them. The metals are then electroplated onto steel wool by passing an electrical current through the solution. Once plated on the steel wool, the precious metals are easily removed by heating in a furnace to about 2000 degrees F[arenheit]. At this temperature, the precious metals melt and separate from the steel. This molten mixture of silver and gold is then poured into a 1000-Troy-ounce mold and allowed to slowly cool. Once chilled, the slag (mixed flux and impurities) that coats the top of the new gold/silver brick is chipped away and recycled. The brick, called a doré bar, is sampled to determine its fineness, stamped with identifying marks, and readied for later shipment to our parent mine in Lead, South Dakota. Doré bars are not pure, like bullion. They generally contain 50% to 75% gold, with the remainder being silver. Final separation of the metals takes place at Lead, where the refined product is fabricated into bars, ingots, beads, and wires. About half of the refined gold is sold as jewelry. The remainder is sold for industrial, medical, dental, electronic, and other purposes.

[Continue past lime silos]

The large silos to our right are our lime storage facility. High-quality lime is shipped in from sources in Arizona and stored in these silos. The neutralization of acid in our ore is a critical part of the cyanide leaching process, and is necessary in order to operate efficiently and safely. Because of this, large inventories of lime are maintained on site. Each silo holds approximately 400 tons of lime.

[Continue past sample prep and assay laboratory]

We are now passing the assay lab and sample prep facility on our right. Here samples from both the mine pit and process facility are analyzed for their gold content and other essential chemical characteristics. We use the age-old technique known as fire assay, but our lab is also equipped with sophisticated equipment that can detect very minute concentrations of many different elements.

Drive to Mine

Our administration building is in Lake County. As we drive east toward the mine, we are passing into Napa County. The road that we are driving on was constructed by Homestake in accordance with state highway standards to accommodate the increased car and heavy truck traffic into this area. Prior to Homestake's arrival, the road to Lower Lake was almost entirely gravel and quite narrow. Power and phone lines were also extended into this area to serve the mine and its neighbors. The construction of the roadway cost about $1 million per mile over 15 miles! The grassy valley we are passing through is Morgan Valley. On our right is Hunting Creek. This is a fairly remote area by local standards, and wildlife abounds here. Deer, tule elk, black bear, bobcats, cougar, golden and bald eagles, and even roadrunners have been sighted here, and their populations are continually monitored. The lands within the mine's security perimeter serve as sanctuary for many of these species. Homestake has even gone as far as to relocate a community of rare bats (Townsend's Big-Eared Bat) that were endangered by the open pit developments!

Davis Creek Reservoir

[The Davis Creek Reservoir is presently not included as part of the tour due to lack of a suitable overlook. The following description of the reservoir may be included at any point in the discussion of mining or milling activities.]

The Davis Creek Reservoir was constructed by Homestake in order to provide the mine and mill with sufficient water to process the ore. The reservoir is contained by a 105-foot earth-fill dam, and has a capacity of 6,000 acre feet when full. It is stocked with various species of fish, but no fishing by employees or the public is permitted. The fish are there to assist in environmental monitoring, especially for concentrations of elements such as mercury in the food chain. Several reservoirs in the northern Coast Ranges, including Clear Lake and Lake Berryessa, show indications of elevated mercury from both natural and artificial causes. The Davis Creek reservoir is also experiencing this phenomenon. It is our hope that Homestake's ongoing studies here will lead to a better understanding of the processes that cause this condition.

Arrival at Rice Paddies

Our next stop is an overlook of the Site 5 Waste Rock area and the sediment containment ponds known as the rice paddies. Don't let the word "waste" scare you. This is in no way toxic waste, but rather rock that contains little or no gold. To expose the gold ore, buried deep in the ground, we must move large amounts of barren rock. That rock is placed in these waste dumps. Ultimately, this waste area will contain nearly 100 million tons of material taken out of the pit. The five ponds you see at the bottom of the dump are sediment control ponds. They are designed to capture any muddy water that runs off the dump. As this water passes from one pond to the next, the silt particles drop out. When the water is finally discharged, it is clear and presents no hazard to the environment. Because the rock in the dumps comes from deep in the earth, it does not contain sufficient nutrients to support vegetation. This has been anticipated, and topsoil stockpiles have been set aside. As the different portions of the waste dump are filled, they will be coated with this topsoil and revegetated. This process has already begun in some portions of the dumps, and will continue for the life of the mine. In this way, this entire area will be reclaimed. It is our goal to leave the area in a condition that will be as good as or better than it was when we came. This waste rock dump is named Site 5 because it was one of a series of sites originally considered for the facility.

[Passing through Homestake gate on way to Barker Dump]

Mine Entrance and Barker Dump Overlook

We are now approaching the area where mining operations are conducted. To see the mining in action, we are taking a special access road to the top of our newly completed Barker dump. This dump, like the Site 5 facility, is composed of waste rock. The Barker dump was completed in the fall of 1990, and has been encapsulated by soil and revegetated. It contains about 15 million tons of rock! When completed, the Site 5 dump will look similar to this, but on an even larger scale. As we drive up the road toward the overlook, notice the south end of the Site 5 dump (still in progress) to the right. Beyond the dump, you see the Knoxville public lands, which are administered by the Bureau of Land Management. Below us on the left is one of our laydown yards, where extra equipment and parts are stored until needed.

[Stop at overlook]

As you can see, the McLaughlin Mine is an open pit mine rather than an underground operation. Open pit mining is a less expensive and safer method and, in this case, is more suitable to our particular orebody. Below us, and to the left (at about 10 o'clock with respect to north), is the M-1 dam and pond. It serves as a temporary containment for water that is pumped from the pit. We are mining below the water table now, so water must constantly be removed from the lowest bench. Because this water tends to be slightly acidic, we do not discharge it into local streams or reservoirs. Instead, it is put to good use in our milling circuit. The little yellow flags with the black stripes you see to the left of the M-1 pond were placed to define the limits of a population of sensitive plants, known as Fritillaria pluriflora (common name: Adobe Lily). We have reserved space for a number of similar rare plant species populations throughout the area that we control. During the dry months, dust control is another problem that must be dealt with. The large equipment used in the mine tends to create dust. To control this, large water trucks are in constant operation, and speed limits are strictly enforced. The mine is surrounded by dust monitoring stations, like large vacuum cleaners, that test the air at intervals. We keep a close eye on these monitors to be sure that we are not exceeding our permitted dust levels.

Pit Design

We are standing above the southwestern edge of the ultimate pit. To orient yourselves, north is toward Little Blue Ridge, which can be seen through the mined-out gap in the ridge at the north end of the pit. Our orebody has two deep roots: One to the north and one to the south. Because of this, we will ultimately have two pits that will overlap slightly, forming a shape like a figure eight. The South Pit is the one you see to the northeast of us (1 to 3 o'clock with respect to north). The north pit excavation has just begun, so it will be a while before it looks much like a pit. The northernmost pit rim is visible at 11 o'clock with respect to north. Ultimately, the two pits will encompass an area over one mile long and half a mile wide. The deepest portion will be about 600 feet below us (the top of the Barker dump is 1,900 feet above sea level). We have several different pit designs available to us. Each is based on the current price of gold. Currently, we are using a design that will allow us to make a profit with gold in the $300 to $450 per ounce range. If the price of gold rises to $600 per ounce, the pit will be expanded to open up deeper gold deposits. Even with the $600 pit design, we will not mine out all the gold-bearing rock. Some of it is just too deep. These deep zones may someday be mined by underground methods. Studies to determine the feasibility of underground mining here are presently under way.

Geology

Before I discuss how mining is done, let me tell you a bit about the geology of this gold deposit. It can get pretty complicated, but I'll try to keep it simple. A large gold deposit such as this one is a very uncommon thing. It is, you might say, a freak of nature! Gold exists in very small quantities in all types of rock on Earth, but you only get high-grade gold veins under very unusual circumstances. To make a deposit like this, you need to follow a sort of recipe! First, you need underground water and lots of it. Second, you need hot magma (molten rock) to heat the water and add certain chemicals that will dissolve gold. Third, you need some gold to dissolve, either in the magma or in surrounding rock deep in the earth. And fourth, you need to draw all that hot water into one small area near the surface of the Earth, so that the dissolved gold can crystallize and concentrate. It just so happens that all those conditions were met here! I'll tell you a bit about the geology, and how the recipe all came together. The McLaughlin

Deposit is located along an ancient fault zone called the Stony Creek fault. This fault is a boundary between serpentine to the west and sandstone to the east. About 2 million years ago, the Stony Creek fault was active (like the San Andreas fault is today). As the rock on either side of the fault moved, hot magma from deep within the earth rose along weaker zones within the fault plane: Magma is one ingredient in the recipe! This magma heated water trapped within the sandstones on the east side of the fault, and filled this water with a variety of dissolved chemicals: water, the second ingredient in the recipe! As this superheated water, charged with chemicals, rose upward (as all hot liquids do), it scavenged gold, silver, and other metals and minerals from the surrounding rocks (the third ingredient in the recipe), and began to transport them upward to the surface. The magma and fluids tended to rise toward this particular spot probably because local bends in the Stony Creek fault acted like pipelines, diverting them to one point and adding the fourth and final ingredient to our recipe. As the hot water approached the surface, it began to boil. This caused the water to cool and to release certain gases which had helped it carry the gold. At that point, the gold, quartz, and other minerals were forced to come out of the solution and crystallize. Just as hard water tends to form crusts in pipes, so the gold and quartz formed crusts on the walls of fractures in the rocks. These fractures—some small, some large—eventually were choked off by these crusts of quartz and gold. They are the veins we mine today. The hot water, after depositing most of its minerals at depth, continued to flow upward. At the surface, it spouted as geysers and hot springs, and formed mineral terraces like those at Yellowstone. We believe this geyser activity reached its peak about 1 million years ago, and died about 500,000 years ago. Before gold mining began here, some of the ancient hot springs terraces (called sinters) were still preserved, hiding beneath them a fortune in gold!

Mining Operations

Mining is done on levels or benches, each usually 20 feet high. The floor of each bench coincides with an even elevation above sea level. Our highest bench is 2,180. When the pit is complete, we may have benches below 1,300! To begin a bench, we must first drill and blast the rock. To start with, a pattern must be laid out. These are broad areas on a bench, where a grid-like pattern of painted rocks is laid out on the ground using surveying instruments for precise location. Drills move onto this pattern and drill holes where the surveyors have indicated. Holes are usually 25 feet deep. As each hole is drilled, a sample of the rock is taken. This sample is later retrieved and sent to our assay lab to determine its gold content and other important chemical characteristics. When the pattern is complete (usually 200 to 300 holes), it is made ready for blasting. To do this, our contract blasting crew (Alpha Explosives) loads each hole with explosives. We use several types of explosive, the most common being ANFO, a mixture of ammonium nitrate (fertilizer) and fuel oil. ANFO is a very safe explosive to use. In fact, in order for it to explode, it must be set off by a small explosive charge called a booster. A web-like pattern of explosive cord ties all the holes in the blast pattern together, so that they detonate in the proper sequence. This is important, because we just want to shatter the rock, and not throw it all over the place! Usually after each day's shift, when everyone is out of the pit, a pattern is blasted. Sirens sound, warning everyone to keep clear. The blasting crew guards the roads as a safety precaution. The blast is set off by means of a hand-held device connected by several hundred feet of cord to the pattern. The assays of the drillhole cuttings taken from individual blastholes are used by the engineering department to determine which portions of the blasted material are waste and which are ore. The individual blasthole locations are plotted on maps and grouped according to the results of the assay. In this way, we can define the limits of ore and waste. These limits are then located on top of the blasted rock using survey instruments, and are marked by pink ribbon and colored flags. When the entire blasted area has been partitioned off by ribbons and flags, the blasted rock (called broken muck) can be mined. We do this with heavy equipment, including huge front-end loaders, called Caterpillar 992-c's; an O&K 120-c power shovel; a P&H 1550 power shovel; and a fleet of 12 Caterpillar 777 (triple-seven) haul trucks. The loader and shovel operators have a great deal of responsibility. A mistake could cost Homestake thousands of dollars. They are supplied with maps of the ore blocks to be mined, and are told where each block is to be shipped. High-grade ore goes directly to the crusher, while lower grade material is sent to the stockpiles, which are those huge mounds of rock to the northwest (at 10 o'clock). The operators must communicate with each truck driver to ensure that the load of rock goes to the correct destination. As complicated as it may seem, mining proceeds quite smoothly. The stockpiles will continue to grow through the life of the mine. They are composed of various sorts of gold-bearing rock that is either too low in grade to be processed at today's economics, or is scheduled for processing through the mill later. We continue to evaluate our processing techniques to find more-effective methods of recovering the gold from this lower-grade material. It is hoped that by introducing alternative ore processing techniques, most or all of the stockpiled rock will someday be processed. The most recently planned improvement to our ore processing system is a flotation plant, which is scheduled for installation in the near future. This plant will assist in making large portions of the stockpiled material profitable to process at today's economics! The mine operates with three crews on 8-hour shifts, 24 hours a day, 5 days a week. More than 60,000 tons of rock are moved each day! To accomplish this, the equipment is designed to handle large volumes of material. The 777 haul trucks weigh 65 tons empty, and hold between 80 and 100 tons per load. Each can be filled in only a couple of minutes by a loader or shovel. Such enormous equipment doesn't come cheap! 992-c loaders cost $800,000 new. 777 haul trucks cost $700,000 each. And then there are the O&K and P&H shovels at $1.3 million and $1.5 million, respectively! You'll notice that traffic in the mine area is left-hand. This is standard in open-pit mining. Basically, it is for safety and ease of operation. The driver's compartment is on the left side of a 777. In order for him or her to see the outside edge of the haul road, the truck must be in the left-hand lane. You might have noticed I said "him or her." Yes, the crew is composed of both men and women in all aspects of the operation. In fact, for a time, our blasting crew was all female! As long as the job gets done, gender makes no difference.

The large building at 10 o'clock is the McLaughlin Mine truck shop. This building houses the maintenance staff and equipment needed to keep the mining machinery in operation. Every piece of equipment needs regular maintenance, and it is the responsibility of the shop to keep them all running properly. The smooth running of the mine depends on them! The mine operations, engineering, and geology offices are also housed in this building. They are conveniently located here, rather than at the administration building, because of the need for close supervision of the daily mine routine.

[Exit Barker dump] [Stop at grinding area parking lot]

Grinding Area Parking Lot

The grassy hill on the left is one of the topsoil stockpiles. It will be used in the dump reclamation. Some topsoil will have to be shipped in from other locations to complete the project. Notice the small wire screen covers on the soil stockpile and the roadcuts? They are there to protect experimental native plant species. Those species that do well in this soil will be used to quickly revegetate the area upon completion of mining activity.

On our right is the crusher area and mill complex. High-grade ore taken from the pit is placed on the lot pad, a wide area just behind the crusher. Usually, the ore is blended from different parts of the ore deposit in order to keep it all uniform in gold content and chemistry as it is processed. These blended piles of ore, called lots, are crushed in our gyratory crusher. Ore is either dumped in directly from a haul truck, or is fed by a loader from the piles set aside on the lot pad (on the other side of the crusher). The crusher reduces the rock to a manageable size (about 8 inches) to be processed by the mill. After it is crushed, the ore is conveyed up a long boom, called a stacker, and dumped into discrete piles. You will notice there are two stackers: One that moves (a slewing stacker) and one that is fixed. The fixed stacker is a recent addition. It is on the right, and is used to process oxide ore for our special oxide circuit. Oxide ore is crushed only on certain shifts. When regular (refractory) ore is being crushed, the fixed stacker is shut off. The ore from each stacker pile is fed by loader into the grinding circuits. These are housed in the large building in front of us. The mill building contains two independent sets of mills. Each set includes a ball mill and a SAG mill. Conveyor belts transport the ore into the mill building. Ore first enters the SAG mill, a 22-foot diameter by 7-foot-long steel-lined revolving drum powered by a 1,500-hp motor. There it is mixed with water and begins the pulverizing process. SAG stands for semi-autogenous grinding, which means that the ore actually helps pulverize itself. To assist in the grinding, about 50 tons of 4-inch-diameter steel balls are kept in the SAG mill. The ore that comes out is sorted by closed-circuit screens. Any particles larger than 5/16 of an inch are sent back for further time in the SAG mill. Smaller particles pass through the screens into the second mill, the ball mill. It is shaped somewhat differently than the SAG mill. It is 15 feet in diameter and 22 feet long. It is also different in that it is rubber-lined and contains 200 tons of 2½-inch steel balls. With the aid of a 3,000-horsepower motor, the ball mill completes the grinding process. When the ore preparation is finished in this mill, it is ground to 200 mesh, that is, it can pass through a screen containing 200 holes per inch. Screens are not used to sort the particles at this stage, however, because particles this small can't be screen-sorted efficiently. Instead, a cyclone sorter, using centrifugal force, performs the task. The final product is about 80% 200 mesh or smaller—the consistency of talcum powder! The ore is now fine enough to be processed. However, it is over 80% water at this point and it would be expensive to send all that water to the process plant. To deal with this, we use the two large 150-foot-diameter tanks beside the mill, called slurry thickeners (ore mixed with water is called slurry). Chemicals called flocculants are added here to help settle out and thicken the slurry. As the large rake continually agitates the mixture, water is drawn off the top and recycled. The slurry is thickened to 45% solids or about half water. At this point, it is a thick mud, about the consistency of thin cream-of-wheat, and is ready to be sent to the process area. The large tanks to the left of the mill are surge tanks. They store the slurry before it gets sent to the process plant. If the mills have to shut down for maintenance or repairs, the capacity of the surge tanks will compensate for some of the downtime. On the other side of the tanks are the GEHO (pronounced gay-ho) brand pumps. These are the only pumps used to pump the slurry from the mine site to the process facility in Lake County. The pumps were manufactured in the Netherlands. As one might imagine, Dutch pumps are very reliable. After all, most of their country would be flooded without pumps! There are eight GEHO pumps. They pump the slurry through a set of surface and underground 8-inch diameter pipelines nearly five miles to the process area.

[Drive back to Post 1]

Post 1

The Core Shed or Core Library can be seen to the left. The Core Library contains more than 35 miles of core samples from the exploration drilling campaigns at the mine. The Core Shed was the first and only building up here from 1980 to 1982 during the exploration and environmental studies period. It was the base of operations then. Now, the grounds around it are used as a picnic area, with facilities for volleyball, basketball, softball, and general recreation. It is available to the public with advance notice only.

End of Tour

Security Post 1 is the hub of our emergency response system. If anything should go wrong at the mine or mill, Post 1 will be notified. It is staffed by trained emergency medical technicians, capable of dealing with any emergency that may arise. Behind the Post is our ambulance, always kept in prime condition. It's the only 4-wheel drive ambulance in the Tri-County area and is available to other agencies (forestry, fire departments, etc.), if needed.

This concludes our tour of the McLaughlin Mine and facilities. I hope you have enjoyed the tour and perhaps have become a bit more familiar with modern gold mining and its technology. We have tried to make the McLaughlin Mine a model for the industry, and we are proud of the results of our efforts. On behalf of the management and staff at McLaughlin, I want to thank you for your interest. Please, drive safely on your return trip home.

REFERENCES

Albright, Horace M. 1986. "Horace M. Albright Mining Lawyer and Executive, U.S. Potash Company, U.S. Borax, 1933 to 1962," an oral history conducted in 1986 by Eleanor Swent. Regional Oral History Office, The Bancroft Library, University of California, Berkeley. https://archive.org/details/mininglawyerexecooalbrrich/mode/2up.

Baylor University. 2012. "Understanding Oral History: Why Do It?" Workshop on the Web: Introduction to Oral History. Baylor University Institute for Oral History. https://www.baylor.edu/content/services/document.php/66420.pdf.

Becker, George F. 1888. *Geology of the Quicksilver Deposits of the Pacific Slope with an Atlas*. Department of the Interior, Monographs of the US Geological Survey. Washington, DC: Government Printing Office.

Borg, Vincent. 2001. "Barrick Completes Merger with Homestake." Barrick Gold Corporation. https://www.barrick.com/English/news/news-details/2001/Barrick-Completes-Merger-with-Homestake/default.aspx.

Bronson, William, and T. H. Watkins. 1977. *Homestake: The Centennial History of America's Greatest Gold Mine*. San Francisco: Homestake Mining Company.

Conger, Harry M., III. 2000. "Mining Career with ASARCO, Kaiser Steel, Consolidation Coal, Homestake, 1955 to 1995: Junior Engineer to Chairman of the Board," an oral history conducted in 1999 and 2000 by Eleanor Swent. Regional Oral History Office, The Bancroft Library, University of California, Berkeley. https://oac.cdlib.org/view?docId=kt3g500310&brand=oac4&doc.view=entire_text.

"Conveyor Belt." 1984. *Napa Valley Register*, November 3, 1984.

Courtney, Kevin. 1984. "Homestake Protests Taxes." *Napa Valley Register*, November 29, 1984.

———. 1994. "Deep Dark Secrets." *Napa Valley Register*, September 18, 1994.

———. 1996. "Soon All Quiet at McLaughlin." *Napa Valley Register*, April 18, 1996.

Coverdale, Jennifer. 1995. "Bat Flap." *Napa Valley Register*, October 30, 1995.

Davis, Kip. 1974. "Manhattan Mine: Last Trace of an Industry." *Napa Valley Register*, November 23, 1974.

Dunning, Gail E. 2008. "The Challenge Mercury Deposit, Redwood City, San Mateo County, California." *Journal of the Bay Area Mineralogists* 9, no. 1 (January): 1–22.

Enderlin, Elmer. 1995. "The Prospectors." [poem]. In *Treasured Poems of America*. Sistersville, WV: Sparowgrass Poetry Forum.

"Environmental Giant Sylvia McLaughlin of Berkeley Dies at Age 99." 2016. *East Bay Express*, January 20. https://www.eastbayexpress.com/SevenDays/archives/2016/01/20/environmental-giant-sylvia-mclaughlin-of-berkeley-dies-at-age-99.

Ernst, Doug. 1984. "Supervisors Uphold Homestake Approval Napa County." *Napa Valley Register*, January 3.

———. 1986. "Toxic Material Brought to Napa Gold Operation." *Napa Valley Register*, May 8.

"Focus: Vintage Crusher." 1984. *Napa Valley Register*, November 3.

"Gold Needs Ordinance." 1980. *Napa Valley Register*, August 28.

"Gold: Below Expectations." 1986. *Napa Valley Register*, July 25.

"Golden Affair." 1985. *Napa Valley Register*, September 30.

Hazen, Wayne C. 1995. "Plutonium Technology Applied to Mineral Processing; Solvent Extraction; Building Hazen Research; 1940 to 1993," an oral history conducted in 1993 by Eleanor Swent. Regional Oral History Office, The Bancroft Library, University of California, Berkeley. https://archive.org/details/plutonium techoohazerich/page/n5/mode/1up.

Herz, Nathaniel. 1966. *Treatise on Analytical Chemistry*, Part 11, *Analytic Chemistry of Inorganic and Organic Compounds*, vol. 4: *Gold*. New York: John Wiley & Sons.

"Homestake's Area Mine Gets BLM Praise." 1992. *Napa Valley Register*, February 9.

Humphrey, William A. 1996. "Mining Operations and Engineering Executive for Anaconda, Newmont, Homestake 1950 to 1995," an oral history conducted in 1994 and 1995 by Eleanor Swent. Regional Oral History Office, The Bancroft Library, University of California, Berkeley. https://archive.org/details/mininganaconda oohumprich/page/n7/mode/2up.

———. 1996. "In His Honor." 1985. *Napa Valley Register*, September 30.

Ingle, Hugh C., Jr. 2000. "Independent Small Mines Operator, 1948 to 1999; Corona Mine," an oral history conducted in 1992, 1997, and 2000 by Eleanor Swent. Regional Oral History Office, The Bancroft Library, University of California, Berkeley. https://oac.cdlib.org/view?docId=kt796nb3xp&brand=oac4&doc.view =entire_text.

"It's Definite, Gold Mining Here in '84." 1980. *Napa Valley Register*, September 17.

"James W. (Bill) Wilder." 2006. *Lake County Record-Bee*, March 1, 2006. https://www .legacy.com/obituaries/record-bee/obituary.aspx?n=james-w-wilder-bill&pid =16898615.

Johnson, Andrew Scott. 2013. *Mercury and the Making of California: Mining, Landscape, and Race, 1840–1890*. Boulder: University Press of Colorado, Boulder.

Knoxville Mining District (KMD). 1998. *Volume 1: The McLaughlin Gold Mine, Northern California, 1978–1995*. Regional Oral History Office, The Bancroft Library, University of California, Berkeley. https://oac.cdlib.org/view?docId=ktoc6002jp&brand =oac4&doc.view=entire_text.

Knoxville Mining District (KMD). 1999. *Volume 2: The McLaughlin Gold Mine, Northern California, 1978–1997*. Regional Oral History Office, The Bancroft Library, University of California, Berkeley. https://oac.cdlib.org/view?docId=kt1r29n6ph&brand =oac4&doc.view=entire_text.

Knoxville Mining District (KMD). 1998. *Volume 3: The McLaughlin Gold Mine, Northern California, 1978–1995*. Regional Oral History Office, The Bancroft Library, University of California, Berkeley. https://oac.cdlib.org/view?docId=kt9779p21d&brand =oac4&doc.view=entire_text.

Knoxville Mining District (KMD). 1998. *Volume 4: The McLaughlin Gold Mine, Northern California, 1978–1995*. Regional Oral History Office, The Bancroft Library, University of California, Berkeley. https://oac.cdlib.org/view?docId=kt5r29n9np&brand=oac4&doc.view=entire_text.

Knoxville Mining District (KMD). 1999. *Volume 5: The McLaughlin Gold Mine, Northern California, 1978–1997*. Regional Oral History Office, The Bancroft Library, University of California, Berkeley. https://oac.cdlib.org/view?docId=kt238nb0gn&brand=oac4&doc.view=entire_text.

Knoxville Mining District (KMD). 1998. *Volume 6: The McLaughlin Gold Mine, Northern California, 1978–1995*. Regional Oral History Office, The Bancroft Library, University of California, Berkeley. https://oac.cdlib.org/view?docId=kt3h4nb1j3&brand=oac4&doc.view=entire_text.

Knoxville Mining District (KMD). 2000. *Volume 7: The McLaughlin Gold Mine, Northern California*. Regional Oral History Office, The Bancroft Library, University of California, Berkeley. https://oac.cdlib.org/view?docId=kt6489n9wp&brand=oac4&doc.view=entire_text.

Knoxville Mining District (KMD). 1998. *Volume 8: The McLaughlin Gold Mine, Northern California*. Regional Oral History Office, The Bancroft Library, University of California, Berkeley. https://archive.org/stream/knoxvillemining08swenrich/knoxvillemining08swenrich_djvu.txt.

Kravig, Clarence. 1995. "From Geologist to Assistant Manager, 1929–1971," an oral history conducted in 1993 by Eleanor Swent in Homestake Mine Workers, Lead, South Dakota. 1929–1993. Regional Oral History Office, The Bancroft Library, University of California, Berkeley. https://archive.org/stream/homestakemine00swenrich/homestakemine00swenrich_djvu.txt.

Kravig, Harford, and Kenneth Kinghorn. 1995. "Homestake Mine Workers, Lead, South Dakota, 1929–1933: Oral History Transcript." Regional Oral History Office, The Bancroft Library, University of California, Berkeley. https://archive.org/stream/homestakemine00swenrich/homestakemine00swenrich_djvu.txt.

Livermore, John Sealy. 2000. "Prospector, Geologist, Public Resource Advocate: Carlin Mine Discovery, 1961; Nevada Gold Rush, 1970s," an oral history conducted in 1992, 1997, and 2000 by Eleanor Swent and Maurice Fuerstenau. Regional Oral History Office, The Bancroft Library, University of California, Berkeley. https://oac.cdlib.org/view?docId=kt796nb3xp&brand=oac4&doc.view=entire_text.

"Looking for Gold." 1991. *Napa Valley Register*, May 15, 1991.

Mason, Richard. "Homestake Gold Mine or Bust." 1985. *Napa Valley Register*, January 31, 1985.

McLaughlin Mine Annual Monitoring Report July 1, 2000–June 30, 2001, includes Closure Plan & Update to Environmental Monitoring Manual. 2001. Bancroft Library supplement to the oral history. The Bancroft Library, University of California, Berkeley.

McLaughlin Natural Reserve. University of California, Davis. https://naturalreserves.ucdavis.edu/mclaughlin-reserve.

"Mercury." 1996. *Napa Valley Register*, November 18, 1996.

Mitchell, Steven. 2009. *Nuggets to Neutrinos: The Homestake Story*. Bloomington, IN: Xlibris Corporation.

"Napa the Latest Mother Lode?" 1980. *Napa Valley Register*, August 28, 1980.

National Mining Association. n.d. "Home Page." National Mining Association, Washington, DC. www.nma.org.

National Mining Hall of Fame and Museum. 2007. "Merrill, Charles Washington." National Mining Hall of Fame and Museum, Leadville, CO. http://www.mininghalloffame.org/inductee/merrill.

Nolte, Carl. 2003. "End of a Golden Era: Mine on Auction Block / $1 Billion in Bullion Pulled from North State in 17 Years." *San Francisco Chronicle*, January 30, 2003.

Nystrom, Eric C. 2014. *Seeing Underground: Maps, Models, and Ming Engineering in America*. Reno: University of Nevada Press.

Oakeshott, Gordon B. 1988. *The California Division of Mines and Geology, 1948–74*. Berkeley: University of California.

Oxford Universal Dictionary. 1955. *The Oxford Universal Dictionary on Historical Principles*, 3rd ed. Oxford, UK: Clarendon Press.

Palmer, Lyman L. 1881. *History of Napa and Lake Counties, California*. San Francisco: Slocum, Bowen & Co., Publishers.

"Photos." 1984. *Napa Valley Register*, November 3, 1984.

Pierson, Elizabeth D. 1989. "Help for Townsend's Big-Eared Bats in California." *Bat Conservation International* 7, no. 1 (Spring 1989). https://www.batcon.org/article/help-for-townsends-big-eared-bats-in-california/.

Pierucci, Tony. 2015. "Sending the Problem Elsewhere." *Lake County Record-Bee*, December 3, 2015.

Procter & Gamble. n.d. "P&G History: A Legacy of Forward-Thinking." Procter & Gamble, Cincinnati, OH.

Public Resources Code. California Department of Conservation. https://www.conservation.ca.gov/dmr/lawsandregulations.

Purtell, Patrick. 1999. "Maintenance and Management at the McLaughlin Mine, 1985 to 1997," an oral history conducted in 1996 and 1997 by Eleanor Swent. Regional Oral History Office, The Bancroft Library, University of California, Berkeley. https://digitalassets.lib.berkeley.edu/roho/ucb/text/purtell_patrick.pdf.

Rosenblatt, Joseph. 1992. "EIMCO, Pioneer in Underground Mining Machinery and Process Equipment, 1926-1963," an oral history conducted by Eleanor Swent. Regional Oral History Office, The Bancroft Library, University of California, Berkeley. https://archive.org/details/pioneerundergroundooroserich.

Western Mining in the Twentieth Century Oral History Series EIMCo, Pioneer in Underground Mining Machinery and Process Equipment. Interviews conducted by Eleanor Swent in 1991. The Bancroft Library, University of California, Berkeley.

SMARA Code Enforcement. Imperial County Planning & Development Services. http://www.icpds.com/?pid=1064.

Sanford Lab Homestake. n.d. "Welcome to Sanford Lab Homestake Visitor Center." Sanford Lab Homestake, Lead, SD.

Silverthorne, Sean. 1980. "Gold at Knoxville, Possibly Largest Modern Discovery." *Napa Valley Register*, August 28, 1980.

———. 1981a. "Napa Gold Glitters, Homestake Mining Reports Record Year." *Napa Valley Register*, March 26, 1981.

———. 1981b. "Bill Just Shrugs Off His Bonanza." *Napa Valley Register*, April 18, 1981.

———. 1981c. "He Hit Rock Bottom Before Gold." *Napa Valley Register*, April 18, 1981.
———. 1986. "Gold Mine Production Tumbles." *Napa Valley Register*, July 25, 1986.
Spratt, Meg. 1988. "Another Attack of Gold Fever." *Napa Valley Register*, April 9, 1988.
Strauss, Simon. 1986. *Trouble in the Third Kingdom: The Minerals Industry in Transition*. London: Mining Journal Books.
Stumbos, John. 2018. "Ecologist Susan Harrison Elected to National Academy." *UC Davis University News*. May 18, 2018. https://www.ucdavis.edu/news/ecologist-susan-harrison-elected-national-academy/.
Swent, Langan W. 1995. "Working for Safety and Health in Underground Mines; San Luis and Homestake Mining Companies, 1946-1988," in two volumes, an oral history conducted in 1987–1988, 1994 by Malca Chall. Regional Oral History Office, The Bancroft Library, University of California, Berkeley. https://archive.org/details/safetyinundergro01swenrich/page/n7/mode/1up.
Taft Law. 2016. "Government's War-Time Closure of Gold Mine Insufficient to Establish 'Operator' Liability Under CERCLA." Taft Law, Cincinnati, OH.
"Terry Mudder is the first inductee for the International Mining Technology Hall of Fame – the event to be held February 22, 2014." 2013. https://im-mining.com/2013/10/11/terry-mudder-is-the-first-inductee-for-the-international-mining-technology-hall-of-fame-the-event-to-be-held-february-22-2014/.
"There's Gold in Them Thar Hills." 1984. *Napa Valley Register*, November 3, 1984.
"Transformation at Knoxville." 1985. *Napa Valley Register*, May 31, 1985.
University of California, Berkeley (UC Berkeley). n.d. "1906 Earthquake." Berkeley Seismology Lab, University of California, Berkeley.
U.S. Geological Survey (USGS). n.d. "Volcano Hazards Program: U.S. Volcanoes and Current Activity Alerts." U.S. Geological Survey, Reston, VA.
UtahRails.net. 2015. https://utahrails.net/mining/new-park-mining.php.
"WASTE: Homestake Hauling Startles County Planners." 1986. *Napa Valley Register*, May 8, 1986.
"Water Diversion for Mine Gets OK." 1984. *Napa Valley Register*, March 16, 1984.
White House. n.d. "Lou Henry Hoover." White House, Washington, DC. www.whitehouse.gov/about-the-white-house/first-ladies/lou-henry-hoover.
Wilder, James William. 1996. "Owner of the Shot Mining Company, Manhattan Mercury Mine, 1951 to 1981," an oral history conducted in 1994 and 1995 by Eleanor Swent." Regional Oral History Office, The Bancroft Library, University of California, Berkeley. https://archive.org/details/ownershotminingoowildrich/page/n5/mode/2up.
Wilson, Alexander. 2000. "Leading a Changing Utah Construction and Mining Company: Utah International, GE-Utah, BHP-Utah 1954 to 1987," an oral history conducted in 1996 and 1997 by Eleanor Swent. Regional Oral History Office, The Bancroft Library, University of California, Berkeley. https://archive.org/details/leadingchangingutahoowilsrich/page/n7/mode/1up.
Woodside, Peter. 1985. "Vintner Strikes Gold (Fleck) in the Vineyard." *Napa Valley Register*, September 26, 1985.
Zuckerman, Seth. 1998. "Homestake Shows How Good a Mine Can be." "The Reclamation of the McLaughlin Mine Received National Acclaim." *High Country News*, January 19, 1998. https://www.hcn.org/issues/122/3885.

INDEX

ABOUT THE AUTHOR

ELEANOR HERZ SWENT was born in Lead, South Dakota, in 1924. Her mother had a degree in geology and her father was chief metallurgist at the Homestake mine. She graduated from Lead High School, was editor of the school newspaper, and president of South Dakota High School Press Association. She attended Northwestern University Journalism Institute and Dana Hall School.

She attended Wellesley College, where she was president of the Shakespeare Society and a member of Phi Beta Kappa. She spent one summer volunteering at a day school for children of workers at the Willow Run defense plant in Michigan, and another on the Pine Ridge and Rosebud Reservations, gathering material for her honors thesis on Sioux Indians. After graduating in 1945 with departmental honors in English, she was employed as assistant to the president of Elmira College in Elmira, New York, in charge of publicity and public relations.

She earned a master of arts in English from the University of Denver in 1947, and married Langan W. Swent, a mining engineer and son of a mining engineer. From 1947 to 1954 the Swents resided in Tayoltita, Durango, Mexico, where he was assistant manager of the San Luis Mine. Three of their children were born during this time; Eleanor traveled to the United States for each delivery.

From 1954 to 1957 they resided in Lead, South Dakota, where he was assistant to the manager of the Homestake mine. In 1957 they moved to Grants, New Mexico, a boomtown where he was manager of Homestake's uranium mining operations. Their fourth child was born in July 1957. Eleanor secured a notary public license, registered more than 2,000 new voters, and was named chairman of a new precinct. She wrote a book review column for the Grants *Daily Beacon* and helped to organize an auxiliary for the new hospital.

In 1966 Langan Swent became vice president of Homestake and they moved to Piedmont, California. She first volunteered and later was employed (1967–86) as a teacher of English as a second language in Oakland

Adult School and local community colleges. In 2010 she published a book based on that experience, *Asian Refugees in America: Narratives of Escape and Adaptation*. She wrote a “Cousin Jenny” column for the California Mining Association newsletter and was awarded their Clementine Award for her contribution to mining history.

In 1986 she became a research interviewer/editor at the Regional Oral History Office, now the Oral History Center, at the Bancroft Library, UC Berkeley, and director of the Western Mining in the Twentieth Century oral history series. The thirteen-volume Knoxville/McLaughlin Mine oral history project is part of this series. She retired in 2005, after completing 63 volumes of oral histories.

In 1998 she was awarded the honorary doctor of letters degree by South Dakota School of Mines and Technology. She served as president of the Mining History Association, and in 2004 received the Rodman Paul award for outstanding contribution to mining history. In July 2005 she gave the keynote address, ”Yanks and Aussies: A Symbiosis,” at the eleventh Australian Mining History Conference in Bendigo.

In October 2005 she retired and moved to a retirement home in Palo Alto, California.